SIP

Understanding the Session Initiation Protocol

For a complete listing of the *Artech House Telecommunications Library*, turn to the back of this book.

SIP

Understanding the Session Initiation Protocol

Alan B. Johnston

Artech House
Boston • London
www.artechhouse.com

Library of Congress Cataloging-in-Publication Data
Johnston, Alan.
    SIP: Understanding the Session Initiation Protocol/ Alan Johnston.
        p.   cm. – (Artech House telecommunications library)
    Includes bibliographical references and index.
    ISBN 1-58053-168-7 (alk. paper)
    1. Computer network protocols.   I. Title   II. Series.

    TK5105.55 .J64   2000
    004.6'2—dc21   00-050817
                                                                             CIP

British Library Cataloguing in Publication Data
Johnston, Alan
    SIP: Understanding the Session Initiation Protocol. — (Artech House
    telecommunications library)
    1. Computer network protocols
    I. Title
    004.6'2

    ISBN 1-58053-168-7

Cover design by Lisa Johnston

© 2001 ARTECH HOUSE, INC.
685 Canton Street
Norwood, MA 02062

International Standard Book Number: 1-58053-168-7
Library of Congress Catalog Card Number: 00-050817

10 9 8 7 6 5 4 3

For Lisa

Contents

Foreword

The Internet now challenges the close to $1 trillion world telecom industry. A renaissance in communications is taking place on the Internet. At its source are new communication protocols that would be impractical on the centralized control systems of circuit-switched networks used in telecommunications.

The Internet and the World Wide Web can be technically defined only by their protocols. Similarly, IP telephony and the wider family of IP communications are defined by several key protocols, most notably by the Session Initiation Protocol, or SIP.

The previously closed door of telecommunications is now wide open to web developers because of SIP and its relation to the web HTTP 1.1 protocol and the e-mail SMTP protocol. IP communications include voice/video, presence, instant messaging, mobility, conferencing, and even games. We believe many other communication areas are yet to be invented. The integration of all types of communications on the Internet may represent the next "killer application" and generate yet another wave of Internet growth.

As explained in this book, SIP is a close relative of the HTTP 1.1 and SMTP protocols. This represents a revolution in communications because it abandons the telecom signaling and control models developed for telephony over many years in favor of Internet and web-based protocols. Users and service providers obtain not only seamless integration of telephony and conferencing with many other World Wide Web and messaging applications, but also benefit from new forms of communications, such as presence and instant messaging.

Mobility can also be managed across various networks and devices using SIP. Location management is now under user control, so that incoming "calls" can be routed to any network and device that the called party may prefer. Users may even move across the globe to another service provider and maintain not only their URL "number", but also their personal tailored services and preferences. The end user gains control over all possible preferences, depending on various parameters such as who the other party is, what network he is on and what devices he is using, as well as time of day, subject, and other variables.

The new dimension in communications called "presence" enables users for the first time to indulge in "polite calling" by first sensing presence and preferences of the other party, before making a call. In its turn, presence can trigger location- and time-dependent user preferences. Users may want to be contacted in different ways, depending on their location and type of network access.

E-commerce will also benefit from IP communications. Extremely complex telecom applications, as found in call centers, have become even more complex when integrated with e-mail and web applications for e-commerce. Such applications, however, are quite straightforward to implement using SIP, due to its common structure with the web and e-mail. For example, both call routing and e-mail routing to agents—based on various criteria such as queue length, skill set, time of day, customer ID, the web page the customer is looking at, and customer history—can be reduced to simple XML scripts when using SIP and another IETF standard, the Call Processing Language (CPL). These examples are in no way exhaustive, but are mentioned here as a way of introduction.

This book starts with a short summary of the Internet, the World Wide Web, and its core protocols and addressing. Though familiar to many readers, these chapters provide useful focus on issues for the topics ahead. The introduction to SIP is made easy and understandable by examples that illustrate the protocol architecture and message details. Finally, in the core of the book, a methodical and complete explanation of SIP is provided. We refer the reader to the Table of Contents for a better overview and navigation through the topics.

Alan Johnston has made significant contributions toward the use of SIP for communications over the Internet. I had the privilege of watching Alan in meetings with some of the largest telecom vendors as he went methodically line by line over hundreds of call flows, which were then submitted as an Internet Draft to the Internet Engineering Task Force (IETF) and implemented in commercial systems. Alan combines in this book his expertise and methodical approach with page turning narrative and a discreet sense of humor.

I could not help reading the book manuscript page by page, since everything from Internet basics, protocols, and SIP itself is explained so well, in an attractive and concise manner.

Henry Sinnreich
Distinguished Member of Engineering
WorldCom
Richardson, Texas
July 2000

Preface

When I began looking into the Session Initiation Protocol (SIP) in October 1998, I had prepared a list of a half dozen protocols relating to Voice over IP and Next Generation Networking. It was only a few days into my study that my list narrowed to just one: SIP. My background was in telecommunications, so I was familiar with the complex suite of protocols used for signaling in the Public Switched Telephone Network. It was readily apparent to me that SIP would be revolutionary in the telecommunications industry. Only a few weeks later I remember describing SIP to a colleague as the "SS7 of future telephony"—quite a bold statement for a protocol that almost no one had heard of, and that was not even yet a proposed standard!

Nearly 2 years later, I have continued to work almost exclusively with SIP since that day in my position with WorldCom, giving seminars and teaching the protocol to others. This book grew out of those seminars and my work on various Internet-Drafts.

This revolutionary protocol was also the discovery of a radical standards body—the Internet Engineering Task Force (IETF). Later, I attended my first IETF meeting, which was for me a career changing event. To interact with this dedicated band of engineers and developers, who have quietly taken the Internet from obscurity into one of the most important technological developments of the late 20th century, for the first time was truly exciting.

Just a few short years later, SIP has taken the telecommunications industry by storm. The industry press contains announcement after announcement of SIP product and service support from established vendors startups, and from established carriers. As each new group and company joins the dialog, the protocol has been able to adapt and grow without becoming unwieldy or overly complex. In the future, I believe that SIP, along with a TCP/IP stack, will find its way into practically every intelligent electronic device that has a need to communicate with the outside world.

With my telecommunications background, it is not surprising that I rely on telephone examples and analogies throughout this book to explain and illustrate SIP. This is also consistent with the probability that telecommunications is the first widely deployed use of the protocol. SIP stacks will soon be in multimedia PCs, laptops, palmtops, and in dedicated SIP telephones. The protocol will be used by telephone switches, gateways, wireless devices, and mobile phones. One of the key features of SIP, however, is its flexibility; as a result, the protocol is likely to be used in a whole host of applications that have little or nothing to do with telephony. Quite possibly one of these applications, such as instant messaging, may become the next "killer application" of the Internet. However, the operation and concepts of the protocol are unchanged regardless of the application, and the telephone analogies and examples are, I feel, easy to follow and comprehend.

The book begins with a discussion of the Internet, the IETF, and the Internet Multimedia Protocol Stack, of which SIP is a part. From there, the protocol is introduced by examples. Next, the elements of a SIP network are discussed, and the details of the protocol in terms of message types, headers, and response codes are covered. In order to make up a complete telephony system, related protocols, including Session Description Protocol (SDP) and Real-Time Transport Protocol (RTP), are covered. SIP is then compared to another signaling protocol, H.323, with the key advantages of SIP highlighted. Finally, the future direction of the evolution of the protocol is examined.

Two of the recurring themes of this book are the simplicity and stateless nature of the protocol. Simplicity is a hallmark of SIP due to its text-encoded, highly readable messages, and its simple transactional model with few exceptions and special conditions. Statelessness relates to the ability of SIP servers to store minimal (or no) information about the state or existence of a media session in a network. The ability of a SIP network to use stateless servers that do not need to record transactions, keep logs, fill and empty buffers, etc., is, I believe, a seminal step in the evolution of communications systems. I hope that these two themes become apparent as you read this book and learn about this exciting new protocol.

The text is filled with examples and sample SIP messages. I had to invent a whole set of IP addresses, domain names, and URLs. Please note that they are *all* fictional—do not try to send anything to them.

I would first like to thank the group of current and former engineers at WorldCom who shared their knowledge of this protocol and gave me the opportunity to author my first Internet-Draft document. I particularly thank Henry Sinnreich, Steve Donovan, Dean Willis, and Matt Cannon. I also thank Robert Sparks, who I first met at the first seminar on SIP that I ever presented. Throughout the whole 3-hour session I kept wondering about the guy with the pony tail who seemed to know more than me about this brand new protocol! Robert and I have spent countless hours discussing fine points of the protocol. In addition, I would like to thank him for his expert review of this manuscript prior to publication—it is a better book due to his thoroughness and attention to detail. I also thank everyone on the IETF SIP list who has assisted me with the protocol and added to my understanding of it.

A special thanks to my wife Lisa for the terrific cover artwork and the cool figures throughout the book.

Finally, I thank my editor Jon Workman, the series editor and reviewer, and the whole team at Artech for helping me in this, my first adventure in publishing.

1

SIP and the Internet

The Session Initiation Protocol (SIP) is a new signaling protocol developed to set up, modify, and tear down multimedia sessions over the Internet [1]. This chapter covers some background for the understanding of the protocol. SIP was developed by the Internet Engineering Task Force (IETF) as part of the Internet Multimedia Conferencing Architecture, and was designed to dovetail with other Internet protocols such as TCP, UDP, IP, DNS, and others. This organization and these related protocols will be briefly introduced. Related background topics such as Internet URLs, IP multicast routing, and ABNF representations of protocol messages will also be covered.

1.1 Signaling Protocols

This book is about the Session Initiation Protocol, which is a signaling protocol. As the name implies, the protocol allows two end-points to establish media sessions with each other. The main functions of signaling protocols are as follows:

- Location of an end-point;
- Contacting end-point to determine willingness to establish a session;
- Exchange of media information to allow session to be established;
- Modification of existing media sessions;

- Tear-down of existing media sessions.

The treatment of SIP in this book will be from a telephony perspective. This is likely to be one of the first applications of SIP, but not the only one. SIP will likely be used to establish a whole set of session types that bear almost no resemblance to a telephone call. The basic protocol operation, however, will be the same. As a result, this book will use familiar telephone examples to illustrate concepts.

1.2   The Internet Engineering Task Force

SIP was developed by the Internet Engineering Task Force (IETF). To quote *The Tao of the IETF* [2]: "The Internet Engineering Task Force is a loosely self-organized group of people who make technical and other contributions to the engineering and evolution of the Internet and its technologies." The two document types used within the IETF are Internet-Drafts (I-Ds) and Request for Comments (RFCs). I-Ds are the working documents of the group; anyone can author one on any topic and submit it to the IETF. There is no formal membership in the IETF; anyone can participate. Every I-D contains the following paragraph on the first page: "Internet-Drafts are documents valid for a maximum of six months and may be updated, replaced, or obsoleted by other documents at any time. It is inappropriate to use Internet-Drafts as reference material or to cite them other than as work in progress."

Internet standards are archived by the IETF as the Request for Comments, or RFC, series of numbered documents. As changes are made in a protocol, or new versions come out, a new RFC document with a new number is issued, which "obsoletes" the old RFC. Some I-Ds are cited in this book; I have tried, however, to restrict this to mature documents that are likely to become RFCs by the time this book is published. A standard begins life as an I-D, then progresses to an RFC once there is consensus and there are working implementations of the protocol. Anyone with Internet access can download any I-D or RFC at no charge using the World Wide Web, ftp, or e-mail. Information on how to do so is on the IETF web site: http://www.ietf.org.

The IETF is organized into working groups, which are chartered to work in a particular area and develop a protocol to solve that particular area. Each working group has its own archive and mailing list, which is where most of the work gets done. The IETF also meets three times per year.

1.3   A Brief History of SIP

SIP was originally developed by the IETF Multi-Party Multimedia Session Control Working Group, known as MMUSIC. Version 1.0 was submitted as an Internet-Draft in 1997. Significant changes were made to the protocol and resulted in a second version, version 2.0, which was submitted as an Internet-Draft in 1998. The protocol achieved Proposed Standard status in March 1999 and was published as RFC 2543 [3] in April 1999. In September 1999, the SIP working group was established by the IETF to meet the growing interest in the protocol. An Internet-Draft containing bug fixes and clarifications to SIP was submitted in July 2000, referred to as RFC 2543 "bis". This document will be first published as an Internet-Draft then as an RFC with a new RFC number, which will obsolete RFC 2543. To advance from Proposed Standard to Draft Standard, a protocol must have multiple independent interworking implementations and limited operational experience. To this end, forums of interoperability tests, called "bakeoffs," have been organized by the SIP working group. Three interoperability "bakeoffs" took place for SIP in 1999, with more planned for 2000. The final level, Standard, is achieved after operational success has been demonstrated [4]. With the documented interoperability of the bakeoffs, SIP should move to Draft Standard status sometime in early 2001.

SIP incorporates elements of two widely used Internet protocols: HTTP (Hyper Text Transport Protocol) used for web browsing and SMTP (Simple Mail Transport Protocol) used for e-mail. From HTTP, SIP borrowed a client-server design and the use of uniform resource locators (URLs). From SMTP, SIP borrowed a text-encoding scheme and header style. For example, SIP reuses SMTP headers such as To, From, Date, and Subject. In keeping with its philosophy of "one problem, one protocol", the IETF designed SIP to be a pure signaling protocol. SIP uses other IETF protocols for transport, media transport, and media description. The interaction of SIP with other Internet protocols such as IP, TCP, UDP, and DNS will be described in the next section.

1.4   Internet Multimedia Protocol Stack

Figure 1.1 shows the four-layer Internet Multimedia Protocol stack. The layers shown and protocols identified will be discussed.

1.4.1   Physical Layer

The lowest layer is the physical and link layer, which could be an Ethernet local area network (LAN), a telephone line (V.90 or 56k modem) running Point-to-Point Protocol (PPP), or a digital subscriber line (DSL) running asynchronous transport mode (ATM), or even a multi-protocol label switching (MPLS) network. This layer performs such functions as symbol exchange, frame synchronization, and physical interface specification.

1.4.2   Internet Layer

The next layer in Figure 1.1 is the Internet layer. Internet Protocol (IP) [5] is used at this layer to route a packet across the network using the destination IP address. IP is a connectionless, best-effort packet delivery protocol. IP packets can be lost, delayed, or received out of sequence. Each packet is routed on its own, using the IP header appended to the physical packet. IP address examples in this book use the current version of IP, Version 4. IPv4 addresses are four octets long, usually written in so-called "dotted decimal" notation (for example, 207.134.3.5). Between each of the dots is a decimal number between 0 and 255. At the IP layer, packets are not acknowledged. A checksum is calculated to detect corruption in the IP header, which could cause a packet to become misrouted. Corruption or errors in the IP *payload*, however, are not detected; a higher layer must perform this function if necessary. IP uses a single octet protocol number in the packet header to identify the transport layer protocol that should receive the packet. IP addresses used over the public Internet are assigned in blocks by the Internet Assigned

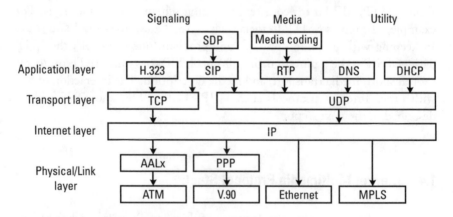

Figure 1.1 The Internet Multimedia Protocol stack.

Number Association (IANA). As a result of this centralized assignment, IP addresses are globally unique. This enables a packet to be routed across the public Internet using only the destination IP address. Various protocols are used to route packets over an IP network, but they are outside of the scope of this book. Subnetting and other aspects of the structure of IP addresses are also not covered here. There are other excellent sources [6] that cover the entire suite of TCP/IP protocols in more detail.

1.4.3   Transport Layer

The next layer shown in Figure 1.1 is the transport layer. It uses a two-octet port number from the application layer to deliver the datagram or segment to the correct application layer protocol at the destination IP address. Some port numbers are dedicated to particular protocols—these ports are called "well-known" port numbers. For example, HTTP uses the well-known port number of 80, while SIP uses the well-known port number of 5060. Other port numbers can be used for any protocol, and they are assigned dynamically from a pool of available numbers. These so-called "ephemeral" port numbers are usually in the range 49152–65535. There are two commonly used transport layer protocols, Transmission Control Protocol (TCP) and User Datagram Protocol (UDP) described in the next sections.

1.4.3.1   TCP

Transmission Control Protocol [7] provides reliable, connection-oriented transport over IP. TCP uses sequence numbers and positive acknowledgments to ensure that each block of data, called a segment, has been received. Lost segments are retransmitted until they are successfully received. Figure 1.2 shows the message exchange to establish and tear down a TCP connection. A TCP server "listens" on a well-known port for a TCP request. The TCP client sends a SYN (synchronization) message to open the connection. The SYN message contains the initial sequence number the client will use during the connection. The server responds with a SYN message containing its own initial sequence number, and an acknowledgment number, indicating that it received the SYN from the client. The client completes the three-way handshake with an ACK or a DATA packet with the AK flag set to the server acknowledging the server's sequence number. Now that the connection is open, either client or server can send data in DATA packets called segments.

Each time a sender transmits a segment, it starts a timer. If a segment is lost in transmission, this timer will expire. Deducing a lost segment, the sender will resend the segment until it receives the acknowledgment. The

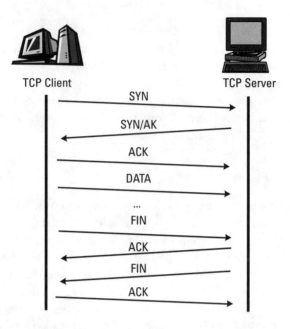

Figure 1.2 Opening and closing a TCP connection.

FIN message closes the TCP connection. The sequence of four messages shown in Figure 1.2 closes the connection. The ephemeral port numbers used in the connection are then free to be used in establishing other connections. TCP also has built-in mechanisms for flow control. During the SYN processing, a window size representing the initial maximum number of unacknowledged segments is sent, which starts at 1 and increases exponentially up to a maximum limit. When network congestion and packet loss occur, the window resets back to 1 and gradually ramps back up again to the maximum limit. A TCP segment header contains 24 octets. Errored segments are detected by a checksum covering both the TCP header and payload.

1.4.3.2 UDP

User Datagram Protocol [8] provides unreliable transport across the Internet. It is a best-effort delivery service, since there is no acknowledgment of sent datagrams. Most of the complexity of TCP is not present, including sequence numbers, acknowledgments, and window sizes. UDP does detect errored datagrams with a checksum. It is up to higher layer protocols, however, to detect this datagram loss and initiate a retransmission if desired.

1.4.4 Application Layer

The top layer shown in Figure 1.1 is the application layer. This includes signaling protocols such as SIP and media transport protocols such as Real-time Transport Protocol (RTP), which is introduced in Section 7.2. Figure 1.1 includes H.323, introduced in Chapter 8, an alternative signaling protocol to SIP developed by the International Telecommunications Union (ITU). Session Description Protocol (SDP), described in Section 7.1, is shown above SIP in the protocol stack because it is carried in a SIP message body. HTTP, SMTP, FTP, and Telnet are all examples of application layer protocols. Because SIP can use any transport protocol, it is shown interacting with both TCP and UDP in Figure 1.1. The use of TCP or UDP transport for SIP will be discussed in the next chapter.

1.5 Utility Applications

Two utility applications are also shown in Figure 1.1 as users of UDP. The most common use of the Domain Name System (DNS, well-known port number 53) is to resolve a symbolic name (such as domain.com, which is easy to remember) into an IP address (which is required by IP to route the packet). Also shown is the Dynamic Host Configuration Protocol (DHCP). DHCP allows an IP device to download configuration information upon initialization. Common fields include a dynamically assigned IP address, DNS addresses, subnet masks, maximum transmission unit (MTU), or maximum packet size, and server addresses for mail and web browsing. Figure 1.3 shows the layer interaction for processing a request. At the top, a URL from the user layer is input to the application layer. URLs, described later in this chapter, are names used to represent resources, hosts, or files on the Internet. The application passes the generated request (for example, a HTTP GET request, which requests a web page download), the URL, and the port number to the transport layer. The transport layer uses a utility to resolve the domain name extracted from the URL into an IP address. The IP address, datagram (or segment, depending on the transport layer protocol used), and protocol number identifying the transport protocol are then passed to the Internet layer. The Internet layer then passes the packet to the physical layer along with a media access control (MAC) address for routing, in the case of a LAN. A response is processed by reversing the above steps. A response received at the physical layer flows back up the layers, with the header information being stripped off and the response data passed upwards towards the user. The main difference is that no utility is used in response processing.

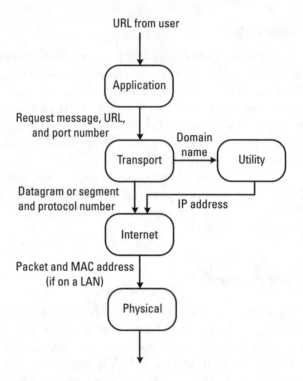

Figure 1.3 Request processing in the Internet Protocol stack.

1.6   DNS and IP Addresses

Domain Name Service [9] is used in the Internet to map a symbolic name (such as www.amazon.com) to an IP address (such as 100.101.102.103). DNS is also used to obtain information needed to route e-mail messages, and, in the future, SIP messages. The use of names instead of numerical addresses is one of the Internet's greatest strengths because it gives the Internet a human, friendly feel. Domain names are organized in a hierarchy. Each level of the name is separated by a *dot*, with the highest level domain on the right side. (Note that the dots in a domain name have no correspondence to the dots in an IP address written in dotted decimal notation.) Top-level domains are shown in Table 1.1. There is also a set of country domains such as: us (United States), uk (United Kingdom), ca (Canada), au (Australia), and so on. Each of these top-level domains has just one authority that assigns that domain to a user or group.

Once a domain name has been assigned, the authority places a link in their DNS server to the DNS server of the user or group who has been assigned the domain. For example, when company.com is allocated to a company, the authoritative DNS server for the top-level com domain entry for company contains the IP address of the company's DNS server(s). A name can then be further qualified by entries in the company's DNS server to point to individual servers in their network. For example, the company's DNS server may contain entries for www.company.com, ftp.company.com, and smtp.company.com. A number of types of DNS record types are defined. The DNS records used to resolve a host name into an IP address are called address records, or A records. Other types of records include CNAME (or canonical name or alias records), MX (or mail exchange records), and TXT(or free-form text records). Another type of DNS record is a PTR, or pointer record, used for reverse lookups. Reverse lookups are used to map an IP address back to a domain name. These records can be used, for example in generating server logs that show not only the IP addresses of clients served, but also their domain name. Web browsing provides an example of the use of the DNS system. When a user types in a web address, such as www.artech-house.com, the name must be resolved to an IP address before the browser can send the request for the index web page from the Artech House web server. The web browser first launches a DNS query to the IP address for its DNS server, which has been manually configured or set up using DHCP. If

Table 1.1
Internet Top-Level Domains

Domain	Description
com	Company
net	Network
int	Internet
org	Not for Profit Organization
edu	University or College
gov	U.S. Government
mil	U.S. Military
arpa	ARPAnet

the DNS server happens to have the name's A record stored locally (cached) from a recent query, it will return the IP address. If not, the DNS root server will then be queried to locate the authoritative DNS server for Artech House, which must contain the A records for the artechhouse.com domain. The HTTP GET request is then sent to that IP address, and the web browsing session begins. There is only one authoritative DNS server for a domain, and it is operated by the owner of the domain name. Due to a very efficient caching scheme built into DNS, a DNS request often does not have to route all the way to this server. DNS is also used by an SMTP server to deliver an e-mail message. An SMTP server with an e-mail message to deliver initiates a DNS request for the MX record of the domain name in the destination e-mail address. The response to the request allows the SMTP server to contact the destination SMTP server and transfer the message. A similar process has been proposed for locating a SIP server using SRV, or service, DNS records.

1.7 URLs

Uniform resource locators [10] are names used to represent addresses or locations in the Internet. URLs are designed to encompass a wide range of protocols and resource types in the Internet. The basic form of a URL is *scheme:specifier*. For example: http://www.artechhouse.com/search/search. html. The token http identifies the scheme or protocol to be used, in this case HTTP. The specifier follows the ":" and contains a domain name (www.artechhouse.com), which can be resolved into an IP address and a file name (/search/search.html). URLs can also contain additional parameters or qualifiers relating to transport. For example telnet://host.company.com:24 indicates that the Telnet Protocol should be used to access host.company.com using port 24. New schemes for URLs for new protocols are easily constructed, and dozens have been defined, such as *mailto, tel, https*, and so on. The SIP URL scheme will be introduced in Section 2.2 and described in detail in Section 4.2.

1.8 Multicast

In normal Internet packet routing, or unicast routing, a packet is routed to a single destination. In multicast routing, a single packet is routed to a set of destinations. Single LAN segments running a protocol such as Ethernet, offer the capability for packet broadcast, where a packet is sent to every node on the network. Scaling this to a larger network with routers is a recipe for disaster, as broadcast traffic can quickly cause congestion. An alternative approach for this type of packet distribution is to use a packet reflector that receives packets and forwards copies to all destinations that are members of a broadcast group. This also can cause congestion in the form of a "packet storm" [11]. For a number of years, the Internet MBONE, or Multicast Backbone Network, an overlay of the public Internet, has used multicast routing for high-bandwidth broadcast sessions. Participants who wish to join a multicast session send a request to join the session to their local MBONE router. That router will then begin to broadcast the multicast session on that LAN segment. Additional requests to join the session from others in the same LAN segment will result in no additional multicast packets being sent, since the packets are already being broadcast. If the router is not aware of any multicast participants on its segment, it will not forward any of the packets. Routing of multicast packets between routers uses special multicast routing protocols to ensure that packet traffic on the backbone is kept to a minimum. Multicast Internet addresses are reserved in the range 224.0.0.0 to 239.255.255.255. Multicast transport is always UDP, since the handshake and acknowledgments of TCP are not possible. Certain addresses have been defined for certain protocols and applications. The scope or extent of a multicast session can be limited using the time-to-live (TTL) field in the IP header. This field is decremented by each router that forwards the packet, limiting the number of hops the packet takes. SIP support for multicast will be discussed in Section 3.8. Multicast is slowly becoming a part of the public Internet as service providers begin supporting it.

1.9 ABNF Representation

The meta-language Augmented Backus-Naur Format (ABNF) [12] is used throughout RFC 2543 [13] to describe the syntax of SIP, as well as other Internet protocols. An example construct used to describe a SIP message is as follows:

```
SIP-message = Request | Response
```

This is read: A SIP message is either a request or a response. SIP-message on the left side of the "equals" sign represents what is being defined. The right side of the "equals" sign contains the definition. The "|" is used to mean *logical* OR (note: "/" is used in place of "|" in some ABNF grammars). Next, `Request` and `Response` are defined in a similar manner using ABNF:

```
Request = Request-Line *( general-header|request-header|
    entity-header ) CRLF [ message-body ] ; Comment
```

`Request-Line` will be defined in another ABNF statement. Elements enclosed in ( ) are treated as a single element. The "*" means the element may be repeated, separated by at least one space. The minimum and maximum numbers can be represented as x*y, which means a minimum of x and maximum of y. Since the default values are 0 and infinity, a solitary "*" (as in this example) indicates any number is allowed, including none. CRLF is defined as a carriage return line feed, or the ASCII characters that are written in Internet hexadecimal notation as *0x10* and *0x13*. Other common ABNF representations include SP for space (ASCII *0x32*). A message body is optional in a Request, and is enclosed in square brackets [ ] to indicate this. Comments in the ABNF begin with a semicolon ";" and continue to the end of the line. Lines continue the same ABNF definition when they are indented. Tokens are defined in ABNF as any set of characters besides control characters and separators. Display names and other components of a SIP header that are not used by the protocol are considered tokens; they are simply parsed and ignored. For example:

```
List = "List"  :  #1 token
```

The "#1" indicates at least one token separated by commas. In this text, few references to ABNF will be made. Instead, SIP messages and elements will be introduced by description and example rather than by using ABNF.

References

[1] Leiner, B., et al., "A Brief History of the Internet," The Internet Society, http://www.isoc.org/internet/history/brief.html.

[2] Malkin, G. and the IETF Secretariat, "The Tao of the IETF—A Guide for New Attendees of the Internet Engineering Task Force," http://www.ietf.org/tao.html.

[3] Handley, M., et al., "SIP: Session Initiation Protocol," RFC 2543, 1999.

[4] Bradner, S., "The Internet Standards Process: Revision 3," RFC 2026, 1996.

[5] "Internet Protocol," RFC 791, 1981.

[6] Wilder, F., *A Guide to the TCP/IP Protocol Suite,* Norwood, MA: Artech House, 1998.

[7] "Transmision Control Protocol," RFC 793, 1981.

[8] Postal, J., "User Datagram Protocol," RFC 768, 1980.

[9] Manning, B., "DNS NSPA RRs," RFC 1348, 1992.

[10] Berners-Lee, T., L. Masintes, and M. McCahill, "Uniform Resource Locators," RFC 1738, 1994.

[11] Hersent, O., D. Gurle, and J. Petit, *IP Telephony Packet-based Multimedia Communications Systems,* Harlow, England: Addison-Wesley, 2000, Chapter 8.

[12] Crocker, D., "Standard for the Format of ARPA Internet Text Messages," RFC 822, 1982.

[13] Handley, M., et al., "SIP: Session Initiation Protocol", RFC 2543, 1999, Appendix C.

2

Introduction to SIP

Often the best way to learn a protocol is to look at examples of its use. While the terminology, structures, and format of a new protocol can be confusing at first read, an example message flow can give a quick grasp of some of the key concepts of a protocol. The example message exchanges in this chapter will introduce SIP.

The first example shows the basic message exchange between two SIP devices. The second example shows the message exchange when a SIP proxy server is used. The third example shows SIP registration. The chapter concludes with a discussion of SIP message transmission using UDP and TCP.

The examples will be introduced using call flow diagrams between a called and calling party, along with the details of each message. Each arrow in the figures represents a SIP message, with the arrowhead indicating the direction of transmission. The thick lines in the figures indicate the media stream. In these examples, the media will be assumed to be Real-timeTrasnport Protocol (RTP) [1] packets containing audio, but it could be another protocol. Details of RTP are covered in Section 7.2.

2.1 A Simple SIP Example

Figure 2.1 shows the SIP message exchange between two SIP-enabled devices. The two devices could be SIP phones, hand-helds, palmtops, or cell

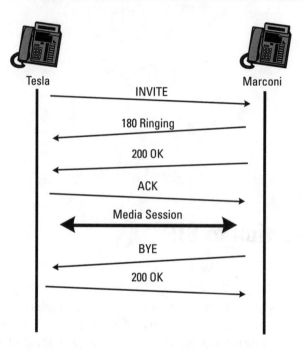

Tesla   Marconi

INVITE

180 Ringing

200 OK

ACK

Media Session

BYE

200 OK

Figure 2.1 A simple SIP example.

phones. It is assumed that both devices are connected to an IP network such as the Internet and know each other's IP address.

The calling party, Tesla, begins the message exchange by sending a SIP INVITE message to the called party, Marconi. The INVITE contains the details of the type of session or call that is requested. It could be a simple voice (audio) session, a multimedia session such as a video conference, or it could be a gaming session.

The INVITE message contains the following fields:

```
INVITE sip:marconi@radio.org SIP/2.0
Via: SIP/2.0/UDP lab.high-voltage.org:5060
To: G. Marconi <sip:Marconi@radio.org>
From: Nikola Tesla <sip:n.tesla@high-voltage.org>
Call-ID: 123456789@lab.high-voltage.org
CSeq: 1 INVITE
Subject: About That Power Outage...
Contact: sip:n.tesla@high-voltage.org
Content-Type: application/sdp
Content-Length: 158
```

```
v=0
o=Tesla 2890844526 2890844526 IN IP4 lab.high-voltage.org
s=Phone Call
c=IN IP4 100.101.102.103
t=0 0
m=audio 49170 RTP/AVP 0
a=rtpmap:0 PCMU/8000
```

The fields listed in the INVITE message are called headers. They have the form Header: Value CRLF. The first line of the request message, called the start line, lists the method, which is INVITE, the Request-URI (Uniform Resource Indicator), then the SIP version number (2.0), all separated by spaces. Each line of a SIP message is terminated by a CRLF. The Request-URI is a special form of SIP URL and indicates the resource to which the request is being sent. SIP URLs and URIs are discussed further in Sections 2.2 and 4.2, respectively.

The first header following the start line is a Via header. Each SIP device that originates or forwards a SIP message stamps its own address in a Via header, usually written as a host name that can be resolved into an IP address using a DNS query. The Via header contains the SIP Version number (2.0), a "/", then UDP for UDP transport, a space, then the hostname or address, a colon, then a port number, in this example the "well-known" SIP port number 5060. Transport of SIP using TCP, UDP, and port numbers are covered later in this chapter.

The next headers are the To and From headers, which show the originator and destination of the SIP request. When a name label is used, as in this example, the SIP URL is enclosed in brackets and used for routing the request. The name can be displayed during alerting.

The Call-ID header has the same form as an e-mail address but is actually an identifier used to keep track of a particular SIP session. The originator of the request creates a locally unique string, then usually adds an "@" and its host name to make it globally unique. The combination of the local address (From header), remote address (To header), and Call-ID identifies the "call leg." The call leg is used by both parties to identify this call because they could have multiple calls set up between them. Subsequent requests for this call will refer to this call leg.

The next header shown is the CSeq, or command sequence. It contains a number, followed by the method name, INVITE in this case. This number is incremented for each new request sent. In this example, the command sequence number is initialized to 1, but it could start at another value.

The Via headers plus the To, From, Call-ID, and CSeq headers represent the minimum required header set in any SIP message. Other headers can be included as optional additional information, or information needed for a specific request type. A Contact header is included in this message, which contains the SIP URL of Tesla; this URL can be used to route messages directly to Tesla. The optional Subject header is present in this example. It is not used by the protocol, but could be displayed during alerting to aid the called party in deciding whether to accept the call. The same sort of useful prioritization and screening we all routinely do using the Subject and From headers in an e-mail message is also possible with a SIP INVITE request. Additional headers are present in this INVITE message, which contain the media information necessary to set up the call.

The Content-Type and Content-Length headers indicate that the message body is Session Description Protocol (SDP) [2] and contains 158 octets of data[1]. A blank line separates message body from the header list, which ends with the Content-Length header. In this case, there are seven lines of SDP data describing the media attributes that the caller Tesla desires for the call. This media information is needed because SIP makes no assumptions about the type of media session to be established—the caller must specify exactly what type of session (audio, video, gaming) that he wishes to establish. The SDP field names are listed in Table 2.1, and will be discussed detail in Section 7.1, but a quick review of the lines shows the basic information necessary to establish a session. This includes the:

- connection IP address (100.101.102.103);
- media format (audio);

---

1. The basis for the octet count of 158 is shown in the following table, where the CR LF at the end of each line is shown as a ©® and the octet count for each line is shown on the right-hand side:

	LINE TOTAL
v=0©®	05
o=Tesla 2890844526 2890844526 IN IP4 lab.high-voltage.org©®	59
s=Phone Call©®	14
c=IN IP4 100.101.102.103©®	26
t=0 0©®	07
m=audio 49170 RTP/AVP 0©®	25
a=rtpmap:0 PCMU/8000©®	22
	Total 158

Table 2.1
SDP Data

SDP parameter	Parameter name
v=0	Version number
o=Tesla 2890844526 2890844526 IN IP4 lab.high-voltage.org	Origin containing name
s=Phone Call	Subject
c=IN IP4 100.101.102.103	Connection
t=0 0	Time
m=audio 49170 RTP/AVP 0	Media
a=rtpmap:0 PCMU/8000	Attributes

- port number (49170);

- media transport protocol (RTP);

- media encoding (PCM μ Law);

- sampling rate (8000 Hz).

INVITE is an example of a SIP request message. There are five other methods or types of SIP requests currently defined in the SIP specification[2]. The next message in Figure 2.1 is a 180 Ringing message sent in response to the INVITE. This message indicates that the called party Marconi has received the INVITE and that alerting is taking place. The alerting could be ringing a phone, flashing a message on a screen, or any other method of attracting the attention of the called party, Marconi.

---

2. Six methods are defined in the base specification RFC 2543. The two other methods described in this text are still in Internet-Draft stage; they are, however, SIP working group items and will shortly be assigned their own RFC numbers.

The 180 Ringing is an example of a SIP response message. Responses are numerical and are classified by the first digit of the number. A 180 response is an "informational class" response, identified by the first digit being a 1. Informational responses are used to convey non-critical information about the progress of the call. SIP response codes were based on HTTP version 1.1 response codes with some extensions and additions. Anyone who has ever browsed the World Wide Web has likely received a "404 Not Found" response from a web server when a requested page was not found. 404 Not Found is also a valid SIP "client error class" response in a call to an unknown user. The other classes of SIP responses are covered in Chapter 5.

Response code number in SIP alone determines the way the response is interpreted by the server or the user. The reason phrase, Ringing in this case, is suggested in the standard, but any text can be used to convey more information. For instance, 180 Hold your horses, I m trying to wake him up! is a perfectly valid SIP response.

The 180 Ringing response has the following structure:

```
SIP/2.0 180 Ringing
Via: SIP/2.0/UDP lab.high-voltage.org:5060
To: G. Marconi <sip:marconi@radio.org>
From: Nikola Tesla <sip:n.tesla@high-voltage.org>
Call-ID: 123456789@lab.high-voltage.org
CSeq: 1 INVITE
Content-Length: 0
```

The message was created by copying many of the headers from the INVITE message, including the Via, To, From, Call-ID, and CSeq, then adding a response start line containing the SIP version number, the response code, and the reason phrase. This approach simplifies the message processing for responses.

Note that the To and From headers are not reversed in the response message as one might expect them to be. Even though this message is sent to Marconi from Tesla, the headers read the opposite. This is because the To and From headers in SIP are defined to indicate the direction of the request, not the direction of the message. Since Tesla initiated this request, all messages will read To: Marconi From: Tesla.

When the called party decides to accept the call (i.e., the phone is answered), a 200 OK response is sent. This response also indicates that the type of media session proposed by the caller is acceptable. The 200 OK is an example of a "success class" response. The 200 OK message body contains Marconi's media information:

```
SIP/2.0 200 OK
Via: SIP/2.0/UDP lab.high-voltage.org:5060
To: G. Marconi <sip:marconi@radio.org>
From: Nikola Tesla <sip:n.tesla@high-voltage.org>
Call-ID: 123456789@lab.high-voltage.org
CSeq: 1 INVITE
Contact: sip:marconi@radio.org
Content-Type: application/sdp
Content-Length: 155

v=0
o=Marconi 2890844528 2890844528 IN IP4 tower.radio.org
s=Phone Call
c=IN IP4 200.201.202.203
t=0 0
m=audio 60000 RTP/AVP 0
a=rtpmap:0 PCMU/8000
```

This response is constructed the same way as the 180 Ringing response. The media capabilities, however, must be communicated in a SDP message body added to the response. From the same SDP fields as Table 2.1, the SDP contains:

- end-point IP address (200.201.202.203);

- media format (audio);

- port number (60000);

- media transport protocol (RTP);

- media encoding (PCM μ Law);

- sampling rate (8000 Hz).

The final step is to confirm the media session with an "acknowledgment" request. The confirmation means that Tesla can support the media session proposed by Marconi. This exchange of media information allows the media session to be established using another protocol, RTP in this example.

```
ACK sip:marconi@radio.org SIP/2.0
Via: SIP/2.0/UDP lab.high-voltage.org:5060
To: G. Marconi <sip:marconi@radio.org>
From: Nikola Tesla <sip:n.tesla@high-voltage.org>
Call-ID: 123456789@lab.high-voltage.org
CSeq: 1 ACK
Content-Length: 0
```

The command sequence, CSeq, has the same number as the INVITE, but the method is set to ACK.

At this point, the media session begins using the media information carried in the SIP messages. The media session takes place using another protocol, typically RTP.

This message exchange shows that SIP is an end-to-end signaling protocol. A SIP network, or SIP server is not required for the protocol to be used. Two end-points running a SIP protocol stack and knowing each other's IP addresses can use SIP to set up a media session between them.

Although less obvious, this example also shows the client-server nature of the SIP protocol. When Tesla originates the INVITE request, he is acting as a SIP client. When Marconi responds to the request, he is acting as a SIP server. After the media session is established, Marconi originates the BYE request and acts as the SIP client, while Tesla acts as the SIP server when he responds. This is why a SIP-enabled device must contain both SIP server and SIP client software—during a typical session, both are needed. This is quite different from other client-server Internet protocols such as HTTP or FTP. The web browser is always an HTTP client, and the web server is always an HTTP server, and similarly for FTP. In SIP, an end-point will switch back and forth during a session between being a client and a server.

In Figure 2.1, a BYE request is sent by Marconi to terminate the media session:

```
BYE sip:n.tesla@high-voltage.org SIP/2.0
Via: SIP/2.0/UDP tower.radio.org:5060
To: Nikola Tesla <sip:n.tesla@high-voltage.org>
From: G. Marconi <sip:marconi@radio.org>
Call-ID: 123456789@lab.high-voltage.org
CSeq: 1 BYE
Content-Length: 0
```

The Via header in this example is populated with Marconi's host address. The To and From headers reflect that this request is originated by Marconi, as they are reversed from the messages in the previous transaction. Tesla, however, is able to identify the call leg and tear down the correct media session.

The confirmation response to the BYE is a 200 OK:

```
SIP/2.0 200 OK
Via: SIP/2.0/UDP tower.radio.org:5060
```

```
To: Nikola Tesla <sip:n.tesla@high-voltage.org>
From: G. Marconi <sip:marconi@radio.org>
Call-ID: 123456789@lab.high-voltage.org
CSeq: 1 BYE
Content-Length: 0
```

The response echoes the `CSeq` of the original request: `1 BYE`.

2.2 SIP Call with Proxy Server

In the SIP message exchange of Figure 2.1, Tesla knew the IP address of Marconi and was able to send the `INVITE` directly to that address. This will not be the case in general—an IP address cannot be used like a telephone number. One reason is that IP addresses are often dynamically assigned due to the shortage of IP version 4 addresses. For example, when a PC dials in to an Internet Service Provider (ISP) modem bank, an IP address is assigned using DHCP to the PC from a pool of available addresses allocated to the ISP. For the duration of the session, the IP address does not change, but it is different for each dial-in session. Even for an "always on" Internet connection such as a DSL line, a different IP address can be assigned after each reboot of the PC. Also, an IP address does not uniquely identify a user, but identifies a node on a particular physical IP network. You have one IP address at your office, another at home, and still another when you log on remotely when you travel. Ideally, there would be one address that would identify you wherever you are. In fact, there is an Internet protocol that does exactly that, with e-mail. SMTP uses a host or system independent name (an e-mail address) that does not correspond to a particular IP address. It allows e-mail messages to reach you regardless of what your IP address is and where you are logged on to the Internet.

SIP uses e-mail-like names for addresses. SIP uses URLs like most Internet protocols. SIP URLs can also handle telephone numbers, transport parameters, and a number of other items. A full description, including examples, can be found in Section 4.2. For now, the key point is that a SIP URL is a name that is resolved to an IP address by using SIP proxy server and DNS lookups at the time of the call, as will be seen in the next example.

Figure 2.2 shows an example of a more typical SIP call with a type of SIP server called a "proxy server." In this example, the caller Schroedinger calls Heisenberg through a SIP proxy server. A SIP proxy operates in a similar way to a proxy in HTTP and other Internet protocols. A SIP proxy does not

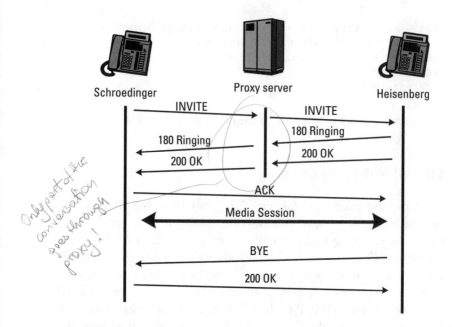

Only part of the conversation goes through proxy!

Figure 2.2 SIP call example with proxy server.

set up or terminate sessions, but sits in the middle of a SIP message exchange, receiving messages and forwarding them. This example shows one proxy, but there can be multiple proxies in a signaling path.

Because Schroedinger does not know exactly where Heisenberg is currently logged on, a SIP proxy server is used to route the INVITE. First, a DNS lookup of Heisenberg's SIP URL domain name (munich.de) is performed, which returns the IP address of the proxy server proxy.munich.de, which handles that domain. The INVITE is then sent to that IP address:

```
INVITE sip:werner.heisenberg@munich.de SIP/2.0
Via: SIP/2.0/UDP 100.101.102.103:5060
To: Heisenberg <sip:werner.heisenberg@munich.de>
From: E. Schroedinger <sip:schroed5244@aol.com>
Call-ID: 10@100.101.102.103
CSeq: 1 INVITE
Subject: Where are you exactly?
Contact: sip:schroed5244@aol.com
Content-Type: application/sdp
Content-Length: 159
```

```
v=0
o=schroed5244 2890844526 2890844526 IN IP4 100.101.102.103
s=Phone Call
t=0 0
c=IN IP4 100.101.102.103
m=audio 49170 RTP/AVP 0
a=rtpmap:0 PCMU/8000
```

The proxy looks up the SIP URL in the `Request-URI` (sip:werner.heisenberg@munich.de) in its database and locates Heisenberg. This completes the two-step process:

- DNS lookup by user agent to locate the IP address of the proxy; Database lookup is performed by the proxy to locate the IP address.

- The `INVITE` is then forwarded to Heisenberg's IP address with the addition of a second `Via` header stamped with the address of the proxy:

```
INVITE sip:werner.heisenberg@200.201.202.203 SIP/2.0
Via: SIP/2.0/UDP proxy.munich.de:5060;branch=83842.1
Via: SIP/2.0/UDP 100.101.102.103:5060
To: Heisenberg <sip:werner.heisenberg@munich.de>
From: E. Schroedinger <sip:schroed5244@aol.com>
Call-ID: 10@100.101.102.103
CSeq: 1 INVITE
Contact: sip:schroed5244@aol.com
Content-Type: application/sdp
Content-Length: 159

v=0
o=schroed5244 2890844526 2890844526 IN IP4 100.101.102.103
s=Phone Call
c=IN IP4 100.101.102.103
t=0 0
m=audio 49172 RTP/AVP 0
a=rtpmap:0 PCMU/8000
```

From the presence of two `Via` headers, Heisenberg knows that the `INVITE` has been routed through a proxy server. The `180 Ringing` response is sent by Heisenberg to the proxy:

```
SIP/2.0 180 Ringing
Via: SIP/2.0/UDP proxy.munich.de:5060;branch=83842.1
Via: SIP/2.0/UDP 100.101.102.103:5060
```

```
To: Heisenberg <sip:werner.heisenberg@munich.de>;tag=314159
From: E. Schroedinger <sip:schroed5244@aol.com>
Call-ID: 10@100.101.102.103
CSeq: 1 INVITE
Content-Length: 0
```

Again, this response contains the Via headers, and the To, From, Call-ID, and CSeq headers from the INVITE request. The response is then sent to the address in the first Via header, proxy.munich.de to the port number listed in the Via header: 5060, in this case. Notice that the To header now has a tag added to it to identify this particular call leg. In more complicated examples, it is possible that a single INVITE can be "forked" and be sent to multiple locations simultaneously. The only way responses from multiple places can be identified (as opposed to a retransmission of a single response) is by the different *tags* on the To headers of the responses.

The proxy receives the response, checks that the first Via header has its own address (proxy.munich.de), removes that Via header, then forwards the response to the address in the next Via header: IP address 100.101.102.103, port 5060. The resulting response sent by the proxy to Schroedinger is:

```
SIP/2.0 180 Ringing
Via: SIP/2.0/UDP 100.101.102.103:5060
To: Heisenberg <sip:werner.heisenberg@munich.de>;tag=314159
From: E. Schroedinger <sip:schroed5244@aol.com>
Call-ID: 10@100.101.102.103
CSeq: 1 INVITE
Content-Length: 0
```

The use of Via headers in routing and forwarding SIP messages reduces complexity in message forwarding. The request required a database lookup by the proxy to be routed. The response requires no lookup because the routing is imbedded in the message in the Via headers. Also, this ensures that responses route back through the same set of proxies as the request.

The call is accepted by Heisenberg, who sends a 200 OK response:

```
SIP/2.0 200 OK
Via: SIP/2.0/UDP proxy.munich.de:5060;branch=83842.1
Via: SIP/2.0/UDP 100.101.102.103:5060
To: Heisenberg <sip:werner.heisenberg@munich.de>;
   tag=314159
From: E. Schroedinger <sip:schroed5244@aol.com>
Call-ID: 10@100.101.102.103
```

```
CSeq: 1 INVITE
Contact: sip:werner.heisenberg@200.201.202.203
Content-Type: application/sdp
Content-Length: 159

v=0
o=heisenberg 2890844526 2890844526 IN IP4 200.201.202.203
s=Phone Call
c=IN IP4 200.201.202.203
t=0 0
m=audio 49172 RTP/AVP 0
a=rtpmap:0 PCMU/8000
```

The proxy forwards the 200 OK message to Schroedinger after removing the first Via header:

```
SIP/2.0 200 OK
Via: SIP/2.0/UDP 100.101.102.103:5060
To: Heisenberg <sip:werner.heisenberg@munich.de>;tag=314159
From: E. Schroedinger <sip:schroed5244@aol.com>
Call-ID: 10@100.101.102.103
CSeq: 1 INVITE
Contact: sip:werner.heisenberg@200.201.202.203
Content-Type: application/sdp
Content-Length: 159

v=0
o=heisenberg 2890844526 2890844526 IN IP4 200.201.202.203
c=IN IP4 200.201.202.203
t=0 0
m=audio 49170 RTP/AVP 0
a=rtpmap:0 PCMU/8000
```

The presence of the Contact header with the SIP URL address of Heisenberg in the 200 OK allows Schroedinger to send the ACK directly to Heisenberg bypassing the proxy. This request, and all future requests continue to use the *tag* in the To header:

```
ACK sip:werner.heisenberg@200.201.202.203 SIP/2.0
Via: SIP/2.0/UDP 100.101.102.103:5060
To: Heisenberg <sip:werner.heisenberg@munich.de>;tag=314159
From: E. Schroedinger <sip:schroed5244@aol.com>
Call-ID: 10@100.101.102.103
```

```
CSeq: 1 ACK
Content-Length: 0
```

This shows that the proxy server is not really "in the call." It facilitates the two end-points locating and contacting each other, but it can drop out of the signaling path as soon as it no longer adds any value to the exchange. A proxy server can force further messaging to route through it by inserting a Record-Route header, which is described in Section 6.2.12. In addition, it is possible to have a proxy server that does not retain any knowledge of the fact that there is a session established between Schroedinger and Heisenberg (referred to as "call state information"). This is discussed in Section 3.3.1.

Note that the media is always end-to-end and not through the proxy. In SIP the path of the signaling messages is totally independent of the path of the media. In telephony, this is described as the separation of control channel and bearer channel.

The media session is ended when Heisenberg sends a BYE message:

```
BYE sip:schroed5244@aol.com SIP/2.0
Via: SIP/2.0/UDP 200.201.202.203:5060
To: E. Schroedinger <sip:schroed5244@aol.com>
From: Heisenberg <sip:werner.heisenberg@munich.de>;
      tag=314159
Call-ID: 10@100.101.102.103
CSeq: 2000 BYE
Content-Length: 0
```

Note that Heisenberg's CSeq was initialized to 2000. Each SIP device maintains its own independent CSeq number space. This is explained in some detail in Section 6.1.3. Schroedinger confirms with a 200 OK response:

```
SIP/2.0 200 OK
Via: SIP/2.0/UDP 200.201.202.203:5060
To: E. Schroedinger <sip:schroed5244@aol.com>
From: Heisenberg <sip:werner.heisenberg@munich.de>;
      tag=314159
Call-ID: 10@100.101.102.103
CSeq: 2000 BYE
Content-Length: 0
```

2.3 SIP Registration Example

Not discussed in the previous example is the question of how the database accessed by the proxy contained Heisenberg's current IP address. There

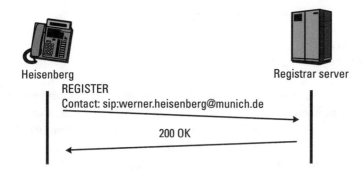

Figure 2.3 SIP registration example.

are many ways this could be done using SIP or other protocols. The mechanism for accomplishing this using SIP is called "registration" and is shown in Figure 2.3.

In this example, Heisenberg sends a SIP REGISTER request to the SIP registrar server. The SIP registrar server receives the message and knows as a result the IP address of Heisenberg. Contained in the REGISTER message is the SIP URL address of Heisenberg. The registrar server stores the SIP URL and the IP address of Heisenberg in a database that can be used, for example, by the proxy server in Figure 2.2 to locate Heisenberg. When a proxy server with access to the database receives an INVITE request addressed to Heisenberg (i.e., an incoming call), the request will be proxied to the stored IP address.

This registration has no real counterpart in the telephone network, but it is very similar to the registration a wireless phone performs when it is turned on. A cell phone sends its identity to the base station (BS), which then forwards the location and phone number of the cell phone to a home location register (HLR). When the mobile switching center (MSC) receives an incoming call, it consults the HLR to get the current location of the cell phone.

The REGISTER message is sent to the SIP registrar server, not to another SIP end device:

```
REGISTER sip:registrar.munich.de SIP/2.0
Via: SIP/2.0/UDP 200.201.202.203:5060
To: Werner Heisenberg <sip:werner.heisenberg@munich.de>
From: Werner Heisenberg <sip:werner.heisenberg@munich.de>
```

```
Call-ID: 23@200.201.202.203
CSeq: 1 REGISTER
Contact: sip:werner.heisenberg@munich.de
Content-Length: 0
```

The Request-URI in the start line of the message contains the address of the registrar server. In a REGISTER request, the To header is the resource that is being registered, in this case sip:werner.heisenberg@munich.de. This results in the To and From headers usually being the same, although an example of third-party registration is given in Section 4.1.2. The SIP URL in the Contact address is stored by the registrar, along with the IP address of Heisenberg.

The registrar server acknowledges the successful registration by sending a 200 OK response to Heisenberg. The response echoes the Contact information that has just been stored in the database:

```
SIP/2.0 200 OK
Via: SIP/2.0/UDP 200.201.202.203:5060
To: Werner Heisenberg <sip:werner.heisenberg@munich.de>
From: Werner Heisenberg <sip:werner.heisenberg@munich.de>
Call-ID: 23@200.201.202.203
CSeq: 1 REGISTER
Contact: sip:werner.heisenberg@munich.de
Content-Length: 0
```

Registration can be automatically performed on initialization of a SIP device.

2.4   Message Transport

As discussed in Chapter 1, SIP is a layer four, or application layer, protocol in the Internet Multimedia Protocol stack shown in Figure 1.1. It can use either TCP or UDP for transport layer, both of which use IP for the Internet layer. How a SIP message is transported using these two protocols will be described in the following sections.

2.4.1   UDP Transport

When using UDP, each SIP request or response message is usually carried by a single UDP datagram or packet. Most SIP messages easily fit in a single datagram. For a particularly large message body, there is a "compact form" of

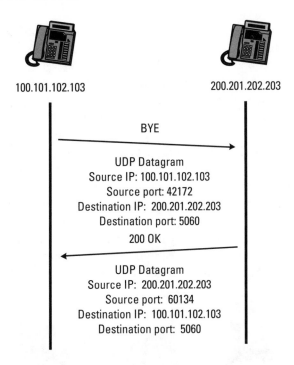

Figure 2.4 Transmission of SIP messages using UDP.

SIP that saves space in representing some headers with a single character. This is discussed in Chapter 6. Figure 2.4 shows a SIP BYE request exchange during an established SIP session using UDP.

The source port is chosen from a pool of available port numbers (above 49172), or sometimes the default SIP port of 5060 is used. The lack of hand-shaking or acknowledgment in UDP transport means that a datagram could be lost and a SIP message along with it. The checksum, however, enables UDP to discard errored datagrams, allowing SIP to assume that a received message is complete and error-free. The reliability mechanisms built into SIP to handle message retransmissions are described in Section 3.5. The reply is also sent to port 5060, or the port number listed in the top via header.

2.4.2 TCP Transport

TCP provides a reliable transport layer, but at a cost of complexity and trans-mission delay over the network. The use of TCP for transport in a SIP message exchange is shown in Figure 2.5. This example shows an INVITE sent by a user agent at 100.101.102.103 to a type of SIP server called a "redirect server" at 200.201.202.203. A SIP redirect server does not forward INVITE requests like a proxy, but looks up the destination address and instead returns that address in a redirection class (3xx) response. The 302 Moved Temporarily response is acknowledged by the user agent with an ACK message. Not shown in this figure is the next step, where the INVITE would be re-sent to the address returned by the redirect server.

As in the UDP example, the "well-known" SIP port number of 5060 is chosen for the destination port, and the source port is chosen from an available pool of port numbers. Before the message can be sent, however, the TCP connection must be opened between the two end-points. This transport layer datagram exchange is shown in Figure 2.5 as a single arrow, but it is actually a three-way handshake between the end-points as shown in Figure 1.2. Once the connection is established, the messages are sent in the stream. The Content-

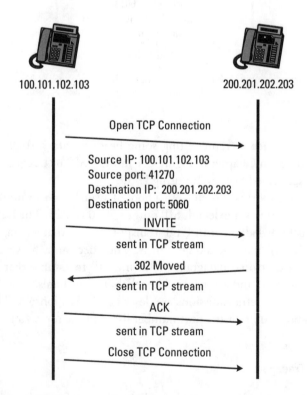

Figure 2.5 Transmission of SIP messages using TCP.

Length header is critical when TCP is used to transport SIP, since it is used to find the end of one message and the start of the next.

The 302 Moved Temporarily response is sent in the stream in the opposite direction. The acknowledgment ACK also is sent in the TCP stream. Because this concludes the SIP session, the connection is then closed. The connection must stay up until the call is established. After that, it can be safely closed without ending the media session. The TCP connection would then need to be reopened to terminate the session with a BYE request.

References

[1] RTP is defined by RFC 1889, "RTP: A Transport Protocol for Real-Time Applications," by H. Schulzrinne, S. Casner, R. Frederick, and V. Jacobson, 1996.

[2] SDP is defined by RFC 2327, "SDP: Session Description Protocol," by M. Handley and V. Jacobson, 1998.

3

SIP Clients and Servers

The client-server nature of SIP has been introduced in the example message flows of Chapter 2. In this chapter, the types of clients and servers in a SIP network will be introduced and defined.

3.1 SIP User Agents

A SIP-enabled end-device is called a SIP user agent (UA). The main purpose of SIP is to enable sessions to be established between user agents. As the name implies, a user agent takes direction or input from a user and acts as an agent on their behalf to set up and tear down media sessions with other user agents. In most cases, the user will be a human, but the user could be another protocol, as in the case of a gateway described in the next section. A user agent must be capable of establishing a media session with another user agent. Since SIP may be used with any transport protocol, there is no requirement that a UA must support either TCP or UDP for message transport. The standard states, however, that a UA should support both TCP and UDP [1].

A UA must maintain state on calls that it initiates or participates in. A minimum call state set includes the local and remote URL, Call-ID, local and remote CSeq headers along with any state information necessary for the media. This information is used to store the call leg and for reliability. The remote CSeq storage is necessary to distinguish between a re-INVITE and a retransmission. A re-INVITE is used to change the session parameters of an existing or pending call. It uses the same Call-ID, but the CSeq is incremented because it is a new request. A retransmitted INVITE will contain the same Call-ID and CSeq as a previous INVITE. Even after a call has been terminated, call state must be maintained by a user agent for at least 32 seconds in case of lost messages in the call tear-down [2].

User agents silently discard an ACK for an unknown call leg. Requests to an unknown URL receive a 404 Not Found Response. A user agent receiving a BYE request for an unknown call leg responds with a 481 Transaction Does Not Exist. Responses from an unknown call leg are also silently discarded. These silent discards are necessary for security. Otherwise, a malicious user agent could gain information about other SIP user agents by spamming fake requests or responses.

A minimum user agent implementation includes support of the methods INVITE and ACK. Although not required to understand every response code defined, a minimal implementation must to be able to interpret any unknown response based on the class (first digit of the number) of the response. That is, if an undefined 498 Wrong Phase of the Moon response is received, it must be treated as a 400 Client Error.

The types of user agents defined in the standard include minimum, basic, redirection, firewall friendly, negotiation, and authentication. These are detailed in Table 3.1. A user agent server responds to an unsupported request with a 501 Not Implemented response.

Most SIP devices support much more than the minimum implementation, and often include support for authentication. A SIP user agent contains both a client application and a server application. The two parts are user agent clients (UAC) and user agent servers (UAS). The UAC initiates requests while the UAS generates responses. During a session, a user agent will usually operate as both a UAC and a UAS.

A SIP user agent must also support SDP for media description. Other types of media descriptions can be used in bodies, but SDP support is mandatory. Details of SDP are in Section 7.1.

3.2 SIP Gateways

A SIP gateway is an application that interfaces a SIP network to a network utilizing another signaling protocol. In terms of the SIP protocol, a gateway is just a special type of user agent, where the user agent acts on behalf of another protocol rather than a human. A gateway terminates the SIP

Table 3.1
User Agent Types

User agent type	Supports
Minimum	INVITE, ACK, SDP, response classes
Basic	Minimum plus BYE
Redirection	Basic plus *Contact* header
Firewall friendly	Redirection plus *Route, Record-Route*, and default proxy server
Negotiation	Firewall plus OPTIONS, *Warning, 380* response
Authentication	Negotiation plus *401* response, *WWW-Authenticate*, and *Authorization* headers

signaling path and can also terminate the media path, although this is not always the case. For example, a SIP to H.323 gateway terminates the SIP signaling path and converts the signaling to H.323, but the SIP user agent and H.323 terminal can exchange RTP media information directly with each other without going through the gateway. An example of this is described in Section 9.6.

A Public Switched Telephone Network (PSTN) gateway terminates both the signaling and media paths. SIP can be translated into, or interwork with, common PSTN protocols such as Integrated Services Digital Network (ISDN), ISDN User Part (ISUP), and other Circuit Associated Signaling (CAS) protocols, which are briefly described in Section 7.4. A PSTN gateway also converts the RTP media stream in the IP network into a standard telephony trunk or line. The conversion of signaling and media paths allows calling to and from the PSTN using SIP. Examples of these gateways are described in Sections 9.3 and 9.4. Figure 3.1 shows a SIP network connected via gateways with the PSTN and a H.323 network.

In the figure, the SIP network, PSTN network, and H.323 networks are shown as "clouds," which obscure the underlying details. Shown connecting to the SIP cloud are SIP IP telephones, SIP-enabled PCs, and corporate SIP gateways with attached telephones. The clouds are connected by gateways. Shown attached to the H.323 network are H.323 terminals and H.323-enabled PCs. The PSTN cloud connects to ordinary analog "black"

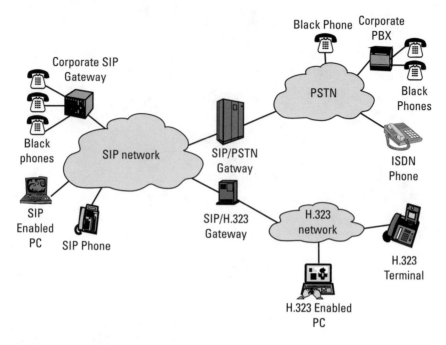

Figure 3.1 SIP network with gateways.

telephones (so called because of the original color of their shell), digital ISDN telephones, and corporate private branch exchanges (PBXs). PBXs connect to the PSTN using shared trunks and provide line interfaces for either analog or digital telephones.

Gateways are sometimes decomposed into a media gateway (MG) and a media gateway controller (MGC). An MGC is sometimes called a "call agent" because it manages call control protocols (signaling), while the MG manages the media connection. This decomposition is transparent to SIP, and the protocols used to decompose a gateway are not described in this book.

Another difference between a user agent and a gateway is the number of users supported. While a user agent typically supports a single user, a gateway can support hundreds or thousands of users. A PSTN gateway could support a large corporate customer, or an entire geographic area. As a result, a gateway does not REGISTER every user it supports in the same way that a user agent might. Instead, a non-SIP protocol can be used to inform proxies about gateways and assist in routing. One protocol that has been proposed for this is the Telephony Routing over IP (TRIP) protocol [3].

3.3 SIP Servers

SIP servers are applications that accept SIP requests and respond to them. A SIP server should not be confused with a user agent server or the client-server nature of the protocol, which describe operation in terms of clients (originators of requests) and servers (originators of responses to requests). A SIP server is a different type of entity. The types of SIP servers discussed in this section are logical entities. Actual SIP server implementations may contain a number of server types, or may operate as a different type of server under different conditions. Because servers provide services and features to user agents, they must support both TCP and UDP for transport. Figure 3.2 shows the interaction of user agents, servers, and a location service. Note that the protocol used between a server and the location service or database is not in general SIP and is not discussed in this book.

3.3.1 Proxy Servers

A SIP proxy server that receives a SIP request from a user agent acts on behalf of the user agent in forwarding or responding to the request. A proxy server

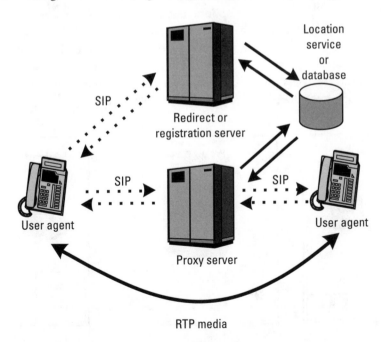

Figure 3.2 SIP user agent, server, and location service interaction.

typically has access to a database or a location service to aid it in processing the request (determining the next hop). The interface between the proxy and the location service is not defined by the SIP protocol. A proxy can use any number of types of databases to aid in processing a request. Databases could contain SIP registrations, or any other type of information about where a user is located. The example of Figure 2.2 introduced a proxy server as a facilitator of SIP message exchange providing user location services to the caller.

A proxy server is different from a user agent or gateway in two key ways:

1. A proxy server does not issue a request; it only responds to requests from a user agent.(A CANCEL request is the only exception to this rule.)

2. A proxy server has no media capabilities.

Figure 3.3 shows the client server interaction of two user agents and a proxy server.

A proxy server can be either stateless or stateful. A stateless proxy server processes each SIP request or response based solely on the message contents. Once the message has been parsed, processed, and forwarded or responded to, no information about the message is stored—no call leg information is stored. A stateless proxy never retransmits a message, and does not use any SIP timers. A stateless proxy has no memory of any requests or responses it has sent or received. A stateless proxy is still capable of detecting message

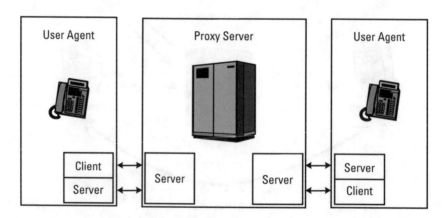

Figure 3.3 User agent and proxy server client/server interaction.

looping since SIP uses a statelesss method to implement loop detection using Via headers.

A stateful proxy server keeps track of requests and responses received in the past and uses that information in processing future requests and responses. For example, a stateful proxy server starts a timer when a request is forwarded. If no response to the request is received within the timer period, the proxy will retransmit the request, relieving the user agent of this task, as described in Section 3.5. Also, a stateful proxy can require user agent authentication, as described in Section 3.6.

A special type of stateful proxy server can receive an INVITE request, then forward it to a number of locations at the same time. This "forking" proxy server keeps track of each of the outstanding requests and the response to each, as shown in Figure 3.4. This is useful if the location service or database lookup returns multiple possible locations for the called party that need to be tried.

A stateful proxy usually sends a 100 Trying response when it receives an INVITE. A stateless proxy never sends a 100 Trying response. A 100 Trying response received by a proxy is never forwarded—it is a single hop only response.

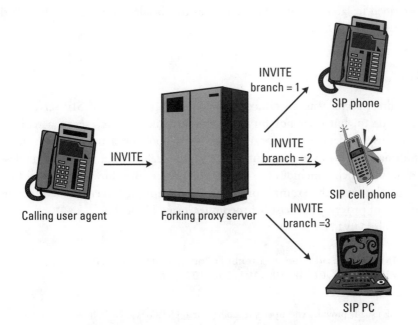

Figure 3.4 Forking proxy operation.

A proxy handling a TCP request must be stateful, since a user agent will assume reliable transport and rely on the proxy for retransmissions on any UDP hops in the signaling path[1].

The only limit to the number of proxies that can forward a message is controlled by the Max-Hops header, which is decremented by each proxy that touches the request. If the Max-Hops count goes to zero, the proxy discards the message and sends a 483 Too Many Hops response back to the originator. Via headers are used to detect message looping. Before forwarding a message, a proxy makes sure its own address is not present in the list of Via headers. If it is and the branch tag (described in Section 6.1.14) matches, then the message has looped and a 482 Loop Detected response is sent after the message is discarded.

A SIP session timer [4] has been proposed to limit the time period over which a stateful proxy must maintain state information. In the initial INVITE request, a Session-Expires header indicates a timer interval after which stateful proxies may discard state information about the session. User agents must tear down the call after the expiration of the timer. The caller can send re-INVITEs to refresh the timer, enabling a "keep alive" mechanism for SIP. This solves the problem of how long to store state information in cases where a BYE request is lost or misdirected, or in other security cases described in later sections. The details of this implementation are described in Section 6.2.19.

3.3.2   Redirect Servers

A redirect server was introduced in Figure 2.5 as a type of SIP server that responds to, but does not forward requests. Like a proxy sever, a redirect server uses a database or location service to look up a user. The location information, however, is sent back to the caller in a redirection class response, which concludes the transaction. Figure 3.5 shows a call flow that is very similar to the example of Figure 2.2, except the server uses redirection instead of proxying to assist Schroedinger locate Heisenberg.

The INVITE contains:

```
INVITE sip:werner.heisenberg@munich.de SIP/2.0
Via: SIP/2.0/UDP 100.101.102.103:5060
```

1. TCP usually provides end-to-end reliability for applications. In SIP, however, TCP only provides single-hop reliability. End-to-end reliability is only achieved by a chain of TCP hops or TCP hops interleaved with UDP hops and stateful proxies.

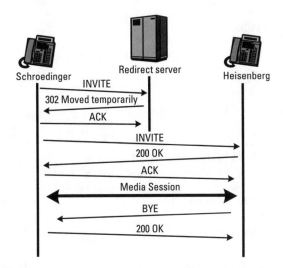

Figure 3.5 Example with redirect server.

```
To: Heisenberg <sip:werner.heisenberg@munich.de>
From: E. Schroedinger <sip:schroed5244@aol.com>
Call-ID: 9@100.101.102.103
CSeq: 1 INVITE
Subject: Where are you exactly?
Contact: sip:schroed5244@aol.com
Content-Type: application/sdp
Content-Length: 159

v=0
o=schroed5244 2890844526 2890844526 IN IP4 100.101.102.103
s=Phone Call
t=0 0
c=IN IP4 100.101.102.103
m=audio 49172 RTP/AVP 0
a=rtpmap:0 PCMU/8000
```

The redirection response to the INVITE is sent by the redirect server:

```
SIP/2.0 302 Moved Temporarily
Via: SIP/2.0/UDP 100.101.102.103:5060
To: Heisenberg <sip:werner.heisenberg@munich.de>;
  tag=052500
From: E. Schroedinger <sip:schroed5244@aol.com>
```

```
Call-ID: 9@100.101.102.103
CSeq: 1 INVITE
Contact: sip:werner.heisenberg@200.201.202.203
Content-Length: 0
```

Schroedinger acknowledges the response:

```
ACK sip:werner.heisenberg@munich.de SIP/2.0
Via: SIP/2.0/UDP 100.101.102.103:5060
To: Heisenberg <sip:werner.heisenberg@munich.de>;tag=052500
From: E. Schroedinger <sip:schroed5244@aol.com>
Call-ID: 9@100.101.102.103
CSeq: 1 ACK
Content-Length: 0
```

This exchange completes this call attempt, so a new INVITE is generated with a new Call-ID and sent directly to the location obtained from the Contact header in the 302 response from the redirect server:

```
INVITE sip:werner.heisenberg@200.201.202.203 SIP/2.0
Via: SIP/2.0/UDP 100.101.102.103:5060
To: Heisenberg <sip:werner.heisenberg@munich.de>
From: E. Schroedinger <sip:schroed5244@aol.com>
Call-ID: 10@100.101.102.103
CSeq: 1 INVITE
Subject: Where are you exactly?
Contact: sip:schroed5244@aol.com
Content-Type: application/sdp
Content-Length: 159

v=0
o=schroed5244 2890844526 2890844526 IN IP4 100.101.102.103
s=Phone Call
t=0 0
c=IN IP4 100.101.102.103
m=audio 49172 RTP/AVP 0
a=rtpmap:0 PCMU/8000
```

The call then proceeds in the same way as Figure 2.2, with the messages being identical. Note that in Figure 3.5, a 180 Ringing response is not sent; instead, the 200 OK response is sent right away. Since 1xx informational responses are optional, this is a perfectly valid response by the UAS if Heisenberg responded to the alerting immediately and accepted the call. In the PSTN, this scenario is called "fast answer."

3.3.3 Registration Servers

A SIP registration server was introduced in the example of Figure 2.3. A registration server accepts SIP REGISTER requests; all other requests receive a 501 Not Implemented response. The contact information from the request is then made available to other SIP servers within the same administrative domain, such as proxies and redirect servers. In a registration request, the To header contains the name of the resource being registered, and the Contact headers contain the alternative addresses or aliases.

Registration servers usually require the registering user agent to be authenticated, using means described in Section 3.6, so that incoming calls can not be hijacked by an unauthorized user. This could be accomplished by an unauthorized user registering someone else's SIP URL to point to their own phone. Incoming calls to that URL would then ring the wrong phone. Depending on the headers present, a REGISTER request can be used by a user agent to retrieve a list of current registrations, clear all registrations, or add a registration URL to the list. These types of requests are described in Section 4.1.2.

3.4 Acknowledgment of Messages

Most SIP requests are end-to-end messages between user agents. That is, proxies between the two user agents simply forward the messages they receive and rely on the user agents to generate acknowledgments or responses.

There are some exceptions to this general rule. The CANCEL method (used to terminate pending calls or searches and discussed in detail in Section 4.1.5) is a hop-by-hop request. A proxy receiving a CANCEL immediately sends a 200 OK response back to the sender and generates a new CANCEL, which is then forwarded to the next hop. (The order of sending the 200 OK and forwarding the CANCEL is not important.) This is shown in Figure 4.4.

Other exceptions to this rule include 4xx, 5xx, and 6xx responses to an INVITE request. While an ACK to a 2xx or 3xx response is generated by the end-point, a 4xx, 5xx, or 6xx response is acknowledged on a hop-by-hop basis. A proxy server receiving one of these responses immediately generates an ACK back to the sender and forwards the response to the next hop. This type of hop-by-hop acknowledgment is shown in Figure 4.2.

ACK messages are only sent to acknowledge responses to INVITE requests. For responses to all other request types, there is no acknowledgment. A lost response is detected by the UAS when the request is retransmitted.

3.5 Reliability

SIP has reliability mechanisms defined, which allow the use of unreliable transport layer protocols such as UDP. When SIP uses TCP, these mechanisms are not used, since it is assumed that TCP will retransmit the message if it is lost and inform the client if the server is unreachable.

For SIP transport using UDP, there is always the possibility of messages being lost or even received out of sequence, because UDP guarantees only that the datagram is error-free. As a result, SIP user agents and servers never perform checks to determine if a message has been corrupted in transmission—it is always assumed to be uncorrupted. It does check, however, to make sure that the UAC has not errored by creating a request missing required headers. Reliability mechanisms in SIP include:

- retransmission timers;
- increasing command sequence CSeq numbers;
- positive acknowledgments.

SIP timer T1 is started by a UAC or a stateful proxy server when a new request is generated or sent. If no response to the request (as identified by a response containing the identical local address, remote address, Call-ID, and CSeq) is received when T1 expires, the request is re-sent. If a provisional (informational class 1xx) response is received, the UAC or stateful proxy server ignores T1 and starts a new longer timer T2. No retransmissions are sent until T2 expires.

After a request is retransmitted, the timer period is doubled until T2 is reached. After that, the remaining retransmissions occur at T2 intervals. This capped exponential backoff process is continued until a maximum of 10 retransmissions at increasing intervals are sent. A stateful proxy server that receives a retransmission of a request discards the retransmission and continues its retransmission schedule based on its own timers. Typically, it will resend the last provisional response.

For INVITE request, the retransmission scheme is slightly different. After a provisional (1XX) response is received, the INVITE is not retransmitted using timer T2.

A stateful proxy must store a forwarded request or generated response message for 32 seconds [2]. After that, only state information need be stored if required. An example message flow involving two user agents, a stateful proxy, two lost messages (shown by "X" in the figure), and three retransmissions is shown in Figure 3.6. In this example, the OPTIONS sent by the stateful proxy server to the UAS is lost. As a result, it is retransmitted when the proxy's T1 timer expires and no response is received. The 200 OK forwarded by the proxy is also lost. When timer T2 in the UAC expires without a final response (non 1xx response), the OPTIONS is retransmitted. When the proxy receives the retransmitted OPTIONS, it deduces that the 200 OK was lost and resends it. The proxy recognizes the 200 OK as a retransmission and does not forward it.

Suggested default values for T1 and T2 are 500ms and 4 seconds, respectively [5]. Longer values are allowed but not shorter ones, because this will generate more message retransmissions.

Note that gaps in CSeq number do not always indicate a lost message. In the authentication examples in the next section, not every request (and hence CSeq) generated by the UAC will reach the UAS if authentication challenges occur by proxies in the path.

3.6 Authentication

Authentication in SIP takes two general forms. One is the authentication of a user agent by a proxy, redirect, or registration server. The other is

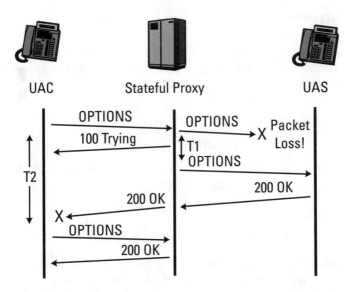

Figure 3.6 SIP reliability example.

the authentication of a user agent by another user agent. A proxy or redirect server might require authentication to allow a user agent to access a service or feature. For example, a proxy server may require authentication before forwarding an INVITE to a gateway or invoking a service. A registration server may require authentication to prevent incoming call hijacking as described previously. User agents can authenticate each other to verify who they are communicating with, since From headers are easily forged[2].

A proxy requiring authentication replies to an unauthenticated INVITE with a 407 Proxy Authorization Required response containing a Proxy-Authenticate header with the form of the challenge. After sending an ACK for the 407, the user agent can then resend the INVITE with a Proxy-Authorization header containing the credentials. User agent, redirect, or registrar servers typically use 401 Unauthorized response to challenge authentication containing a WWW-Authenticate header, and expect the re-INVITE to contain an Authorization header containing the user agent's credentials. A user agent's credentials are usually an encrypted username and password, as in the example of Section 9.1

A call flow involving both proxy and user agent authentication is shown in Figure 3.7.

3.7 Encryption

While authentication is used as a means of access control and identity confirmation, encryption is used for privacy. SIP messages intercepted during session setup reveal considerable information, including:

- Both parties' SIP URLs and IP addresses;
- The fact that the two parties have established a call;
- The IP addresses and port numbers associated with the media, allowing eavesdropping.

---

2. The ability to forge From headers is present in SMTP, where it is virtually a *feature*. A preference setting in an e-mail program sets your name and e-mail address, which need not correspond to the address or domain that is used to send the message. This allows a user to send multiple e-mail addresses from the same e-mail account by simply changing the From address before sending a message. Only a detailed examination of a full set of SMTP headers will show that the e-mail was sent from another address.

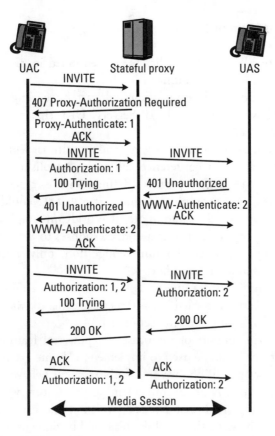

Figure 3.7 Authentication call flow.

SIP supports the encryption of both message bodies and message headers. The encryption of message bodies makes it more difficult for an eavesdropper to listen in. Also, an uninvited third party, knowing all the SDP information could guess the RTP SSRC (page 131) number and send unwanted media to either party, so-called media "spamming". Encryption of headers such as To, From, and Call-ID can hide the parties in a media session. Headers containing information used by proxies to route the requests and responses, however, must not be hidden or the messages will become unroutable. These headers must either be left in the clear, or the proxies must be able to decrypt and encrypt the headers. The details of how this can be done is described in Section 6.1.5 on the Encryption header and Section 6.2.6 on the Hide header.

3.8   Multicast Support

SIP support for UDP multicast has been mentioned in previous sections. There are two main uses for multicast in SIP.

SIP registration can be done using multicast, by sending the REGISTER message to the well-known "All SIP Servers" URL sip:sip.mcast.net at IP address 224.0.1.75.

The second use for multicast is to send a multicast session invitation. This effectively allows a conference call to be established with a single request. An INVITE with a partially defined Request-URI can be sent using multicast. For example, a multicast INVITE could be sent to sip:*@wcom.com, which would invite all WorldCom employees with SIP phones receiving the request to respond. Responses to a multicast request are also sent by multicast. To limit congestion, only a limited set of responses is allowed by the standard.

A proxy can forward a unicast INVITE request to a multicast address. The use of multicast is recorded in a SIP message using the maddr parameter in the via header, as discussed in Section 6.1.14.

This built-in support of multicast is a powerful feature of SIP. For example, multicast could be used to implement a "home extension" feature, which makes a set of SIP phones on a LAN segment behave like telephone extensions at a house. A proxy implementing this feature would convert a unicast INVITE to a multicast INVITE, which would cause all the"extensions" to ring at the same time. While this could be done using only unicast INVITEs and a forking proxy, using multicast is much simpler and more flexible.

Media sessions can be multicast because RTP also supports multicast, allowing conference calling without a conference "bridge." This is not really, however, a feature of SIP, but rather a feature of SDP and RTP, which also support multicast.

3.9   Firewalls and NAT Interaction

Most corporate LANs or intranets connect to the public Internet through a firewall. A firewall is a router that is used to protect the LAN behind it from various kinds of attacks and unauthorized access. Sometimes they are used to prevent users behind the firewall accessing certain resources in the Internet. In the simplest deployment, a firewall can be thought of as a one-way gate—it allows outgoing packets from the intranet to the Internet, but blocks incoming packets from the Internet unless they are responses to queries. Only certain types of requests from the Internet will be allowed to pass through the firewall, such as HTTP requests to the corporate web server, SMTP e-mail messages, or DNS queries to the authoritative DNS for the corporate domain. The firewall does this by keeping track of TCP connections opened and filtering ports.

Firewalls pose a particularly difficult challenge to SIP sessions. Because SIP can use TCP, configuring a firewall to pass SIP is not too difficult. This does not help the media path, however, which uses UDP and will be blocked by most firewalls. A firewall needs to understand SIP, be able to parse an INVITE request, extract the IP addresses and port numbers from the SDP, and open up "pin holes" in the firewall to allow this traffic to pass. The hole can then be closed when a BYE is sent or a session timer expires[3].

Network address translators (NATs) also cause serious problems for SIP. A NAT can be used to conserve IPv4 addresses, or can be used for security purposes, to hide the IP address and LAN structure behind the NAT. It is used on a router or firewall that provides the only connection of a LAN to the Internet, a so-called stub network. A NAT allows non-unique IP addresses to be used internally within the LAN. When a packet is sent from the LAN to the Internet, the NAT changes the non-globally unique address (usually addresses in the range 10.x.x.x, 172.16.x.x - 172.29.x.x and 192.168.x.x) in the packet header to a globally unique address from a pool of available addresses. Addresses can also be statically assigned. This means that every node on the network does not have to have a globally unique IP address. Responses from the Internet are translated back to the non-unique address. A NAT, however, is not completely transparent to higher layers. For a signaling protocol such as SIP, a NAT can cause particular problems.

Because responses in SIP are routed using Via headers, a device behind a NAT will stamp its non-routable private IP address in its Via header of messages that it originates. When the request is forwarded outside the intranet by the NAT, the UDP and IP packet headers will be rewritten with a temporarily assigned global Internet address. The NAT will keep track of the *binding* between the local address and the global address so that incoming packets can have the UDP and IP headers rewritten and routed correctly. However, IP addresses in a SIP message, such as Via headers, or IP addresses in SDP message bodies will not be rewritten and will not be routable. To

---

3. A change to SDP is needed to accomplish this, because both the source and destination IP addresses and port numbers must be contained in the SDP to be visible to the firewall. See Section 7.1 on SDP fields.

solve the message routing problem, SIP has a mechanism for detecting if a NAT is present in a SIP message path. Each proxy or user agent that receives a request checks the received IP address with the address in the v i a header. If the addresses are different, there is a NAT between them. The unroutable v i a header is fixed with a r e c e i v e d tag containing the actual global IP address. Outside the NAT, the response is routed using the received IP address. Inside the NAT, the v i a address is used. This does solve the message response routing problem, but not the media problems.

Another problem with NATs is the time span of the NAT address binding. For a TCP connection, this is not an issue—the binding is maintained as long as the connection is open. For a UDP SIP session, the time period is determined by the application. If a binding were removed before a BYE was sent terminating the session, the connection would effectively be closed and future signaling impossible. For this reason, a NAT-aware SIP proxy server is the best solution to these problems. The proxy would rewrite the media IP addresses in the message setup and would not allow the NAT to remove the address binding until a BYE was sent or a session timer had expired.

References

[1]   Handley, M., et al., "SIP: Session Initiation Protocol," RFC 2543, 1999, Section 1.5.2.

[2]   Handley, M., et al., "SIP: Session Initiation Protocol," RFC 2543, 1999, Section 10.1.2.

[3]   Rosenberg, J., H. Salama, and M. Squire, "Telephony Routing over IP (TRIP)," IETF Internet-Draft, Work in Progress.

[4]   Donovan, S., and J. Rosenberg, "The SIP Session Timer," IETF Internet-Draft, Work in Progress.

[5]   Handley, M., et al., "SIP: Session Initiation Protocol," RFC 2543, 1999, Section 10.4.1.

4

SIP Request Messages

This chapter covers the types of SIP requests called methods. Six are described in the SIP specification document [1]. Two more methods are work items of the SIP working group and will likely achieve RFC status in the near future. Other proposed methods are still in the early stages of development, or have not yet achieved working group consensus (some of these are described in Chapter 10). After discussing the eight methods, this chapter concludes with a discussion of SIP URLs and URIs, tags, and message bodies.

4.1 Methods

SIP requests or methods are considered "verbs" in the protocol, since they request a specific action to be taken by another user agent or proxy server. The INVITE, REGISTER, BYE, ACK, CANCEL, and OPTIONS methods are the original six methods in version 2.0 of SIP. The INFO and PRACK methods are described in separate Internet-Drafts that are likely to become RFCs in the near future.

A proxy does not need to understand a request method in order to forward the request. A proxy treats an unknown method as if it were an OPTIONS; that is, it forwards the request to the destination if it can. This allows new features and methods useful for user agents to be introduced without requiring support from proxies that may be in the middle. A user agent receiving a method it does not support replies with a 501 Not Implemented response. Method names are case sensitive.

4.1.1   INVITE

The INVITE method is used to establish media sessions between user agents. In telephony, it is similar to a setup message in ISDN or an initial address message, or IAM, in ISUP. (PSTN protocols are briefly introduced in Section 7.4.) Responses to INVITEs are always acknowledged with the ACK method described in Section 4.1.4. Examples of the use of the INVITE method are described in Chapter 2.

An INVITE usually has a message body containing the media information of the caller. In addition, it can also contain quality of service (QoS) or security information. A QoS or security message body could carry a token used to authorize the reservation of bandwidth or access to some other resource in the IP network. If an INVITE does not contain media information, the ACK contains the media information of the user agent server (UAS). An example of this call flow is shown in Figure 4.1. If the media information contained in the ACK is not acceptable, then the called party must send a BYE to cancel the session—a CANCEL cannot be sent because the session is already established. A media session is considered established

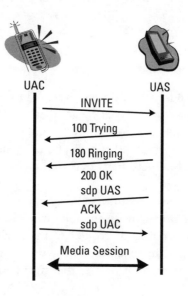

Figure 4.1   INVITE with no SDP message body.

when the INVITE, 200 OK, and ACK messages have been exchanged between the user agent client (UAC) and the UAS. The media session continues until a BYE is sent by either party to end the session, as described in Section 4.1.3.

A UAC that originates an INVITE creates a globally unique Call-ID that is used for the duration of the call. A CSeq count is initialized (which need not be set to 1, but must be an integer) and incremented for each new request for the same Call-ID. The To and From headers are populated with the remote and local addresses. With the exception of the addition of tags, which are described in Section 4.3, the To and From headers cannot be modified or changed during the call.

An INVITE sent for an existing call leg references the same Call-ID as the original INVITE. Sometimes called a re-INVITE, the request is used to change the session characteristics. The CSeq command sequence number is incremented so that a UAS can distinguish the re-INVITE from a retransmission of the original INVITE.

A CANCEL sent with a CSeq that matches a re-INVITE only cancels the change in media session requested—the established media session continues until a BYE is sent by either party. The same is true if the re-INVITE is refused or fails in any way. A re-INVITE should not be sent by a UAC until a final response to the initial INVITE has been received. There is an additional case where two user agents simultaneously send re-INVITEs to each other. This is handled in the same way with a Retry-After header. This condition is called "glare" in telephony, and occurs when both ends of a trunk group seize the same trunk at the same time.

An Expires header in an INVITE indicates to the UAS how long the call request is valid. For example, the UAS could leave an unanswered INVITE request displayed on a screen for the duration of specified in the Expires header. Once a session is established, the Expires header has no meaning—the expiration of the time does not terminate the media session. Instead, a Session-Expires header [2] can be used to place a time limit on an established session.

An example INVITE request with a SDP message body is shown below:

```
INVITE <sip:411@salzburg.aut>;user=phone SIP/2.0
Via: SIP/2.0/UDP salzburg.edu.aut:5060
To: <sip:411@salzburg.aut>;user=phone
From: Christian Doppler <sip:C.doppler@salzburg.edu.aut>
Call-ID: 12-45-A5-46-F5@salzburg.edu.aut
CSeq: 1 INVITE
```

```
Subject: Train Timetables
Contact: sip:c.doppler@salzburg.edu.aut
Content-Type: application/sdp
Content-Length: 152

v=0
o=doppler 2890842326 2890844532 IN IP4 salzburg.edu.aut
s=Phone Call
c=IN IP4 50.61.72.83
t=0 0
m=audio 49172 RTP/AVP 0
a=rtpmap:0 PCMU/8000
```

In addition to the required headers, this request contains the optional Subject header. Note that this Request-URI contains a phone number. Phone number support in SIP URLs is described in Section 4.2.

The mandatory and optional headers in an INVITE request are shown in Table 4.1.

Table 4.1
Mandatory and Optional Headers in an INVITE Request

Mandatory headers	Optional headers	
Call-ID	Accept	Proxy-Authorization
Content-Length	Accept-Encoding	Proxy-Require
CSeq	Accept-Language	Record-Route
From	Authorization	Require
To	Content-Language	Response-Key
Via	Content-Disposition	Route
Contact	Content-Encoding	Server
	Date	Session-Expires
	Encryption	Subject
	Expires	Supported
	Hide	Timestamp
	In-Reply-To	Unsupported
	Max-Forwards	Supported
	MIME-Version	User Agent
	Organization	Warning
	Priority	WWW-Authenticate

4.1.2 REGISTER

The REGISTER method is used by a user agent to notify a SIP network of its current IP address and the URLs for which it would like to receive calls. As mentioned in Section 2.3, SIP registration bears some similarity to cell phone registration on initialization. Registration is not required to enable a user agent to use a proxy server for outgoing calls. It is necessary, however, for a user agent to register to receive incoming calls from proxies that serve that domain unless some non-SIP mechanism is used by the location service to populate the SIP URLs and IP addresses of user agents. A REGISTER request may contain a message body although its use is not defined in the standard [3]. Depending on the use of the Contact and Expires headers in the REGISTER request, the registrar server will take different action. Examples of this are shown in Table 4.2. If no expires parameter or Expires header is present, a SIP URL will expire in 1 hour. The presence of an Expires header sets the expiration of SIP URLs with no expires parameter. If an expires

Table 4.2
Types of Registrar Actions and Contact Headers

Request headers	Registrar action
Contact: * Expires: 0	Cancel all registrations
Contact: sip:galvani@bologna.edu.it; expires=30	Add URL to current registrations; registration expires in 30 minutes
Contact: sip:galvani@bologna.edu.it Expires: 30	Add URL to current registrations; registration expires in 30 minutes
Contact: sip:galvani@bolognauni.edu; expires=45 Contact: sip:l.galvani@bologna.it Expires: 30	Add all URLs to current registrations in preference order listed; first URL expires in 45 minutes, second in 30 minutes

Contact: sip:galvani@bologna.edu.it ;
  action=proxy ;q=0.9
Contact:mailto:galvani@bologna.edu.it;
  q=0.1
	Add URLs to current registrations using specified preference SIP requests should be proxied; SIP URL expires in 60 minutes (default); mailto URL does not expire

No Contact header present	Return all current registrations in response

parameter is present, it sets the expiration time for that Contact only. Non-SIP URLs have no default expiration time.

The CSeq is incremented for a REGISTER request. The use of the Request-URI, To, From, and Call-ID headers in a REGISTER request is slightly different than for other requests. The Request-URI contains only the domain of the registrar server with no user portion. The REGISTER request may be forwarded or proxied until it reaches the authoritative registrar server for the specified domain. The To header contains the SIP URL of the user agent that is being registered. The From contains the SIP URL of the sender of the request, usually the same as the To header. It is recommended that the same Call-ID be used for all registrations by a user agent.

A user agent sending a REGISTER request may receive a 3xx redirection or 4xx failure response containing a Contact header of the location to which registrations should be sent.

A third-party registration occurs when the party sending the registration request is not the party that is being registered. In this case, the From header will contain the URL of the party submitting the registration on behalf of the party identified in the To header. Chapter 3 contains an example of a first-party registration. An example third-party registration request for the user Euclid is shown below:

```
REGISTER sip:registrar.athens.gr SIP/2.0
Via: SIP/2.0/UDP 201.202.203.204:5060
To: sip:euclid@athens.gr
From: sip:secretary@academy.athens.gr
Call-ID: 2000-July-07-23:59:59.1234@201.202.203.204
CSeq: 1 REGISTER
Contact: sip:euclid@parthenon.athens.gr
Contact: mailto:euclid@geometry.org
Content-Length: 0
```

The mandatory and optional headers in a REGISTER request are shown in Table 4.3.

4.1.3   BYE

The BYE method is used to terminate an established media session. In telephony, it is similar to a release message. A session is considered established if an INVITE has received a success class response (2xx) and an ACK has been sent. A BYE is sent only by user agents participating in the session, never by proxies or other third parties. It is an end-to-end method so

Table 4.3

Mandatory and Optional Headers in a REGISTER Request

Mandatory headers	Optional headers	
Call-ID	Accept	Proxy-Authorization
Content-Length	Accept-Encoding	Proxy-Require
CSeq	Accept-Language	Record-Route
From	Authorization	Require
To	Contact	Retry-After
Via	Content-Disposition	Response-Key
	Content-Type	Route
	Content-Encoding	Server
	Date	Supported
	Encryption	Timestamp
	Expires	Unsupported
	Hide	Supported
	Max-Forwards	User Agent
	MIME-Version	Warning
	Organization	WWW-Authenticate
	Proxy-Authenticate	

responses are only generated by the other user agent. A user agent responds with a 486 Unknown Call Leg to a BYE for an unknown call leg.

This method may not contain a message body. A BYE can be sent by either the caller or the called party in a session. A BYE always increments the CSeq. A BYE sent for a pending request cancels the request, but a final response must still be issued for the INVITE by the UAS.

It is not recommended that a BYE be used to cancel pending requests because it will not be forked like an INVITE and may not reach the same set of user agents as the INVITE. An example BYE request looks like the following:

```
BYE sip:info@hypotenuse.org SIP/2.0
Via: SIP/2.0/UDP port443.hotmail.com:5060
To: <sip:info@hypotenuse.org>;tag=63104
From: sip:pythag42@hotmail.com
Call-ID: 34283291273@port443.hotmail.com
```

```
CSeq: 47 BYE
Content-Length: 0
```

The mandatory and optional headers in a BYE request are shown in Table 4.4.

4.1.4 ACK

The ACK method is used to acknowledge final responses to INVITE requests. Final responses to all other requests are never acknowledged. Final responses are defined as 2xx, 3xx, 4xx, 5xx, or 6xx class responses. The CSeq number is never incremented for an ACK, but the CSeq method is changed to ACK. This is so that a UAS can match the CSeq number of the ACK with the number of the corresponding INVITE.

An ACK may contain an application/sdp message body. This is permitted if the initial INVITE did not contain a SDP message body. If the INVITE contained a message body, the ACK may not contain a message

Table 4.4
Mandatory and Optional Headers in a BYE Request

Mandatory headers	Optional headers	
Call-ID	Accept	Proxy-Authorization
Content-Length	Accept-Encoding	Proxy-Require
CSeq	Accept-Language	Record-Route
From	Authorization	Require
To	Content-Disposition	Response-Key
Via	Content-Encoding	Route
	Content-Language	Server
	Content-Type	Supported
	Date	Timestamp
	Encryption	Unsupported
	Expires	Supported
	Hide	User Agent
	Max-Forwards	Warning
	MIME-Version	WWW-Authenticate
	Proxy-Authenticate	

body. The ACK may not be used to modify a media description that has already been sent in the initial INVITE; a re-INVITE must be used for this purpose. This is used in some interworking scenarios with other protocols where the media characteristics may not be known when the initial INVITE is generated and sent. An example of this is described in Section 9.6.

For 2xx responses, the ACK is end-to-end, but for all other final responses it is done on a hop-by-hop basis when stateful proxies are involved. The end-to-end nature of ACKs to 2xx responses allows a message body to be transported. An ACK generated in a hop-by-hop acknowledgment will contain just a single Via header with the address of the proxy server generating the ACK. The difference between hop-by-hop acknowledgments to a response end-to-end acknowledgments is shown in the message fragments of Figure 4.2.

A stateful proxy receiving an ACK message must determine whether or not the ACK should be forwarded downstream to another proxy or user agent or not. That is, is the ACK a hop-by-hop ACK or an end-to-end ACK. This is done by comparing the To (including tags), From, CSeq, and Call-ID headers to those of any non-2xx final responses sent. If there is not an exact match, the ACK is proxied toward the UAS. Otherwise, the ACK is for this hop and is not forwarded by the proxy. The call flows of Chapter 9 show

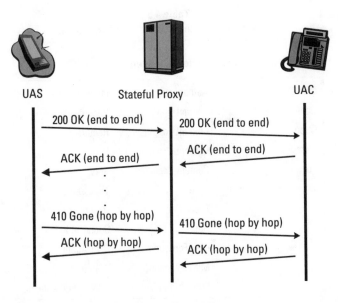

Figure 4.2 End-to-end versus hop-by-hop acknowledgments.

examples of both types of ACK handling. An example ACK containing SDP contains:

```
ACK sip:laplace@mathematica.org SIP/2.0
Via: SIP/2.0/TCP 128.5.2.1:5060
To: Marquis de Laplace <sip:laplace@mathematica.org>;tag=90210
From: Nathaniel Bowditch <sip:n.bowditch@salem.ma.us>
Call-ID: 152-45-N-32-23-W@128.5.2.1
CSeq: 3 ACK
Content-Type: application/sdp
Content-Length: 143

v=0
o=bowditch 2590844326 2590944532 IN IP4 salem.ma.us
s=Bearing
c=IN IP4 128.5.2.1
t=0 0
m=audio 32852 RTP/AVP 0
a=rtpmap:0 PCMU/8000
```

The mandatory and optional headers in an ACK message are shown in Table 4.5.

Table 4.5
Mandatory and Optional Headers in an ACK Request

Mandatory headers	Optional headers	
Call-ID	Authorization	Proxy-Require
Content-Length	Contact	Record-Route
CSeq	Content-Length	Require
From	Content-Disposition	Route
To	Content-Encoding	Server
Via	Date	Timestamp
	Encryption	Unsupported
	Hide	Supported
	Max-Forwards	User Agent
	MIME-Version	Warning
	Proxy-Authorization	WWW-Authenticate

4.1.5   CANCEL

The CANCEL method is used to terminate pending searches or call attempts. It can be generated by either user agents or proxy servers. A user agent uses the method to cancel a pending call attempt it had earlier initiated. A forking proxy can use the method to cancel pending parallel branches after a successful response has been proxied back to the UAC. CANCEL is a hop-by-hop request and receives a response generated by the next stateful element. The difference between a hop-by-hop request and an end-to-end request is shown in Figure 4.3. The CSeq is not incremented for this method so that proxies and user agents can match the CSeq of the CANCEL with the CSeq of the pending INVITE that it corresponds to.

A proxy receiving a CANCEL forwards the CANCEL to the same set of locations with pending requests that the initial INVITE was sent to. A proxy does not wait for responses to the forwarded CANCEL requests, but responds immediately. A user agent confirms the cancellation with a 200 OK response and replies to the INVITE with a 487 Request Canceled response. A CANCEL whose intent is to cancel all pending requests should not contain a tag in the To header, even if provisional responses have been returned containing a tag. This is so that all branches of a forked INVITE will be canceled, not just the one that has returned the provisional response.

If a final response has already been received, a user agent should not generate a CANCEL. Instead, the user agent will need to send an ACK and a

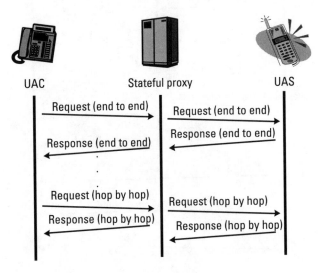

Figure 4.3 End-to-end versus hop-by-hop requests.

BYE to terminate the session. This is also the case in the race condition where a CANCEL and a final response cross in the network, as shown in Figure 4.4. In this example, the CANCEL and 200 OK response messages cross between the proxy and the UAS. The proxy still replies to the CANCEL with a 200 OK, but then also forwards the 200 OK response to the INVITE. The 200 OK response to the CANCEL sent by the proxy only means that the CANCEL request was received and has been forwarded - the UAC must still be prepared to receive further final responses. No 487 response is sent in this scenario. The session is canceled by the UAC sending an ACK then a BYE in response to the 200 OK.

Since it is a hop-by-hop request, a CANCEL may not contain a message body. An example CANCEL request contains:

```
CANCEL sip:i.newton@cambridge.edu.gb SIP/2.0
Via: SIP/2.0/UDP 10.downing.gb:5060
To: Isaac Newton <sip:i.newton@cambridge.edu.gb>
```

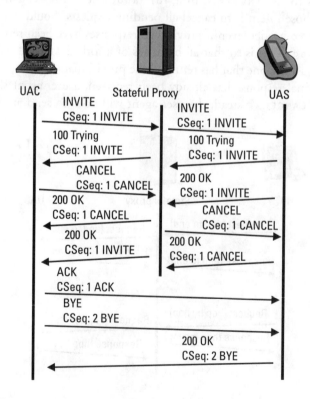

Figure 4.4 Race condition in call cancellation.

```
From: Rene Descartes <sip:visitor@10.downing.gb>
Call-ID: 42@10.downing.gb
CSeq: 32156 CANCEL
Content-Length: 0
```

The mandatory and optional headers in a CANCEL request are shown in Table 4.6.

4.1.6 OPTIONS

The OPTIONS method is used to query a user agent or server about its capabilities and discover its current availability. The response to the request lists the capabilities of the user agent or server. A proxy never generates an OPTIONS request. A user agent or server responds to the request as it would to an INVITE (i.e., if it is not accepting calls, it would respond with a 4xx or 6xx response). A success class (2xx) response can contain Allow, Accept, Accept-Encoding, Accept-Language, and Supported headers indicating its capabilities.

Table 4.6
Mandatory and Optional Headers in a CANCEL Request

Mandatory headers	Optional headers	
Call-ID	Accept	Proxy-Require
Content-Length	Accept-Encoding	Record-Route
CSeq	Accept-Language	Require
From	Authorization	Response-Key
To	Content-Language	Server
Via	Date	Supported
	Encryption	Timestamp
	Hide	Unsupported
	In-Reply-To	Supported
	Max-Forwards	User Agent
	MIME-Version	Warning
	Proxy-Authorization	WWW-Authenticate

An OPTIONS request may not contain a message body. A proxy determines if an OPTIONS request is for itself by examining the Request-URI. If the Request-URI contains the address of the proxy, the request is for the proxy. Otherwise, the options is for another proxy or user agent and the request is forwarded. An example OPTIONS request and response contains:

```
OPTIONS sip:proxy.carrier.com SIP/2.0
Via: SIP/2.0/UDP cavendish.kings.cambridge.edu.uk
To: sip:proxy.carrier.com
From: J.C. Maxwell <sip:james.maxwell@kings.cambridge.edu.uk>
Call-ID: 9352812@cavendish.kings.cambridge.edu.uk
CSeq: 1 OPTIONS
Content-Length: 0

SIP/2.0 200 OK
Via: SIP/2.0/UDP cavendish.kings.cambridge.edu.uk;tag=512A6
To: sip:proxy.carrier.com
From: J.C. Maxwell <sip:james.maxwell@kings.cambridge.edu.uk>
Call-ID: 9352812@cavendish.kings.cambridge.edu.uk
CSeq: 1 OPTIONS
Allow: INVITE, OPTIONS, ACK, BYE, CANCEL, REGISTER
Accept-Language: en, de, fr
Content-Length: 0
```

The mandatory and optional headers in an OPTIONS request is the same as Table 4.3 except Retry-After is not an allowed header.

4.1.7 INFO

The INFO [4] method is used by a user agent to send call signaling information to another user agent with which it has an established media session. This is different from a re-INVITE since it does not change the media characteristics of the call. The request is end to end, and is never initiated by proxies. A proxy will always forward an INFO request—it is up to the UAS to check to see if the call leg is valid. INFO requests for unknown call legs receive a 481 Unknown Call Leg response.

An INFO method typically contains a message body. The contents may be signaling information, a mid-call event, or some sort of stimulus. INFO has been proposed to carry certain PSTN mid-call signaling information such as ISUP USR messages. Section 9.3 contains an example of the use of the INFO method.

The INFO method always increments the CSeq. An example INFO method is:

```
INFO sip:poynting@mason.edu.uk SIP/2.0
Via: SIP/2.0/UDP cavendish.kings.cambridge.edu.uk
To: John Poynting <sip:nting@mason.edu.uk>;tag=432485820183
From: J.C. Maxwell <sip:james.maxwell@kings.cambridge.edu.uk>
Call-ID: 18437@cavendish.kings.cambridge.edu.uk
CSeq: 6 INFO
Content-Type: plain/text
Content-Length: 16

USR.51a6324127
```

The mandatory and optional headers in an INFO request are shown in Table 4.7.

Table 4.7
Mandatory and Optional Headers in an INFO Request

Mandatory headers	Optional headers	
Call-ID	Accept	Priority
Cseq	Accept-Encoding	Proxy-Authorization
From	Accept-Language	Proxy-Require
To	Authorization	Record-Route
Via	Contact	Require
	Content-Length	Response-Key
	Content-Disposition	Route
	Content-Encoding	Server
	Date	Subject
	Encryption	Supported
	Expires	Timestamp
	Hide	Unsupported
	In-Reply-To	Supported
	Max-Forwards	User Agent
	MIME-Version	Warning
	Organization	WWW-Authenticate

4.1.8 PRACK

The PRACK [5] method is used to acknowledge receipt of reliably transported provisional responses (1xx). The reliability of 2xx, 3xx, 4xx, 5xx, and 6xx responses to INVITEs is achieved using the ACK method. However, in cases where a provisional response, such as 180 Ringing, is critical in determining the call state, it may be necessary for the receipt of a provisional response to be confirmed. The PRACK method applies to all provisional responses except the 100 Trying response, which is never reliably transported.

A PRACK is generated by a UAC when a provisional response has been received containing a RSeq reliable sequence number and a Supported: 100rel header. The PRACK echoes the number in the RSeq and the CSeq of the response in a RAck header. The message flow is as shown in Figure 4.5. In this example, the UAC sends the 180 Ringing response reliably by including the RSeq header. When no PRACK is received from the UAC after the expiration of SIP timer T1, the response is retransmitted. The receipt of the PRACK confirms the delivery of the response and stops all further transmissions. The 200 OK response to the PRACK stops retransmissions of the PRACK request. The call completes when the UAC sends the ACK in response to the 200 OK.

Reliable responses are retransmitted using the same exponential backoff mechanism used for final responses to an INVITE. The combination of Call-ID, CSeq number, and RAck number allows the UAC to match the PRACK to the provisional response it is acknowledging. As shown in Figure 4.5, the PRACK receives a 200 OK response, which can be distinguished from the 200 OK to the INVITE by the method contained in the CSeq header. The detailed use of the method is described in Sections 6.2.18 and 6.3.7 (where the RAck and RSeq headers are described) and in Section 9.3 (where a PRACK is used to acknowledge receipt of a 183 Session Progress response in PSTN interworking).

The PRACK method always increments the CSeq. A PRACK may contain a message body. An example exchange contains:

```
SIP/2.0 180 Ringing
Via: SIP/2.0/UDP lucasian.trinity.cambridge.edu.uk
To: Descartes <sip:rene.descartes@metaphysics.org>
From: Newton <sip:newton@kings.cambridge.edu.uk>;tag=981
Call-ID: 5@lucasian.trinity.cambridge.edu.uk
RSeq: 314
CSeq: 1 INVITE
Content-Length: 0
```

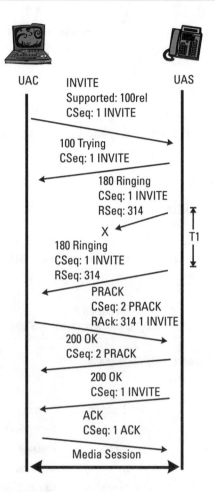

Figure 4.5 Use of reliable provisional responses.

```
PRACK sip:rene.descartes@metaphysics.org SIP/2.0
Via: SIP/2.0/UDP lucasian.trinity.cambridge.edu.uk
To: Descartes <sip:rene.descartes@metaphysics.org>
From: Newton <sip:newton@kings.cambridge.edu.uk>;tag=981
Call-ID: 5@lucasian.trinity.cambridge.edu.uk
CSeq: 2 PRACK
RAck: 314 1 INVITE
Content-Length: 0

SIP/2.0 200 OK
Via: SIP/2.0/UDP lucasian.trinity.cambridge.edu.uk
```

```
To: Descartes <sip:rene.descartes@metaphysics.org>
From: Newton <sip:newton@kings.cambridge.edu.uk>;tag=981
Call-ID: 5@lucasian.trinity.cambridge.edu.uk
CSeq: 2 PRACK
Content-Length: 0
```

The mandatory and optional headers in a PRACK request are shown in Table 4.8.

4.2 SIP URLs and URIs

The addressing scheme of SIP uniform resource locators and uniform resource indicators has been previously mentioned. SIP URLs are used in a

Table 4.8
Mandatory and Optional Headers in a PRACK Request

Mandatory headers	Optional headers	
Call-ID	Accept	Proxy-Authorization
Content-Length	Accept-Encoding	Proxy-Require
CSeq	Accept-Language	Record-Route
From	Authorization	Require
To	Contact	Response-Key
Via	Content-Length	Route
	Content-Disposition	Server
	Content-Encoding	Subject
	Date	Supported
	Encryption	Timestamp
	Expires	Unsupported
	Hide	Supported
	In-Reply-To	User Agent
	Max-Forwards	Warning
	MIME-Version	WWW-Authenticate
	Organization	
	Priority	

number of places including the To, From, and Contact headers, as well as the Request-URI, which indicates the destination. SIP URLs are similar to the mailto URL [6] and can be used in hyperlinks on web pages, for example. SIP URLs can include telephone numbers. The information in a SIP URL indicates the way in which the resource (user) should be contacted using SIP.

An example SIP URL contains the scheme sip a ":", then a user-name@host or IPv4 address followed by an optional ":", then the port number, or a list of ";" separated URI parameters:

```
sip:joseph.fourier@transform.org:5060;transport=udp;user=ip
 ;method=INVITE;ttl=1;maddr=240.101.102.103?Subject=FFT
```

Some SIP URLs, such as a REGISTER Request-URI do not have a username, but begin with the host or IPv4 address. In this example, the port number is shown as 5060, the well-known port number for SIP. If the port number is not present, 5060 is assumed. The transport parameter indicates UDP is to be used, which is the default. TCP is an alternative transport parameter.

The user parameter is used by parsers to determine if a telephone number is present in the username portion of the URL. The assumed default is that it is not, indicated by the value ip. If a telephone number is present, it is indicated by the value phone. This parameter must not be used to guess at the characteristics or capabilities of the user agent. For example, the presence of a user=phone parameter must not be interpreted that the user agent is a SIP telephone (which may have limited display/processing capabilities). In a telephony environment, IP telephones and IP/PSTN gateways may in fact use the reverse assumption, interpreting any digits in a username as digits regardless if user=phone is present.

The method parameter is used to indicate the method to be used. The default is INVITE. This parameter has no meaning in To or From headers, but can be used in Contact headers for registration, for example.

The ttl parameter is the time-to-live, which must only be used if the maddr parameter contains a multicast address and the transport parameter contains udp. The default value is 1. This value scopes the multicast session broadcast, as described in Section 1.8.

The maddr usually contains the multicast address to which the request should be directed, overriding the address in the host portion of the URL. It can also contain, however, a unicast address of an alternative server for requests. The maddr parameter is required for SIP URLs in Record-Route and Route headers.

The method, maddr, ttl, and header parameters must not be included in To or From headers, but may be included in Contact headers or in Request-URIs. In addition to these parameters, a SIP URL may contain other user-defined parameters.

Following the "?" parameter, names can be specified to be included in the request. This is similar to the operation of the mailto URL, which allows Subject and Priority to be set for the request. Additional headers can be specified, separated by a "&". The header name body indicates that the contents of a message body for an INVITE request is being specified in the URL.

If the parameter user=phone is present, then the username portion of the URL can be interpreted as a telephone number. This allows additional parameters in the username portion of the URL. An example username showing these parameters is:

```
#70w555-1212;isub=1000;postd=pp555.1212
```

In this example, the dialed digit string, interpreted by a PSTN gateway, would be the DTMF digit # then 70 (to cancel call waiting, for example), wait for second dial tone w, the digits 555-1212. Additional parameters include an ISDN subaddress of 1000, and a post dial digits of two 1-second pauses pp, then 555.1212. This example shows both types of optional visual separators allowed, either "-"or "."as the separator.

4.3 Tags

A tag is a cryptographically random number with at least 32 bits of randomness, which is added to To and From headers to uniquely identify a call leg. The examples of Chapter 3 and Chapter 9 show the use of the tag header parameter. The To header in the initial INVITE will not contain a tag. If it is possible that the caller may establish multiple sessions with the called party, the caller adds a tag to the From header. Excluding 100 Trying, all responses will have a tag added to the To header. A tag returned in a 200 OK response is then incorporated into the call leg and used in all future requests for this Call-ID. A tag is never copied across calls. Any response generated by a proxy will have a tag added by the proxy.

If a UAC receives responses containing different tags, this means that the responses are from different UASs, and hence the INVITE has been forked. It is up to the UAC as to how to deal with this situation. For example, the UAC could establish separate sessions with each of the responding

UAS. The call legs would contain the same From, Call-ID, and CSeq, but would have different tags in the To header. The UAC also could CANCEL or BYE certain legs and establish only one session.

Since the intent of a CANCEL is to cancel all pending call attempts, regardless of forking, tags are usually not used in the CANCEL, even if they have been received in a provisional response. Individual pending legs of a call cannot be cancelled by sending CANCEL containing the appropriate tag. Note that tags are not part of the URL but are part of the header, and always placed outside any "<>".

4.4 Message Bodies

Message bodies in SIP may contain various types of information. They may contain SDP information, which can be used to convey media information or QoS or even security information.

The optional Content-Disposition header is used to indicate the intended use of the message body. If not present, the function is assumed to be session, which means that the body describes a media session. Besides session, the other defined function is render, which means that the message body should be presented to the user or otherwise used or displayed. This could be used to pass a small JPEG image file or URL.

The format of the message body is indicated by the Content-Type header described in Section 6.4.5. If a message contains a message body, the message must include a Content-Type header. All user agents must support a Content-Type of application/sdp. The encoding scheme of the message body is indicated in the Content-Encoding header. If not specified, the encoding is assumed to be text/plain. The specification of a Content-Encoding scheme allows the message body to be compressed.

The Content-Length header contains the number of octets in the message body. If there is no message body, the Content-Length header should still be included but has a value of 0. Because multiple SIP messages can be sent in a TCP stream, the Content-Length count is a reliable way to detect when one message ends and another begins. If a Content-Length is not present, the UAC must assume that the message body continues until the end of the UDP datagram, or until the TCP connection is closed, depending on the transport protocol.

Message bodies can have multiple parts if they are encoded using Multipart Internet Mail Extensions (MIME) [7]. Message bodies in SIP, however, should be small enough so that they do not exceed the UDP MTU of the network. Proxies may reject requests with large message bodies with a `413 Request Entity Too Large` response, since processing large messages can load a server.

References

[1]   Handley, M., et al., "SIP: Session Initiation Protocol," RFC 2543, 1999.

[2]   Donovan, S., and J. Rosenberg, "The SIP Session Timer," IETF Internet-Draft, Work in Progress.

[3]   Handley, M., et al., "SIP: Session Initiation Protocol," RFC 2543, 1999. Section 8.1, Body Inclusion, states that the use of message bodies in REGISTER requests is for further study. It has been proposed that this could be used by a user agent to upload a call processing language script to run a feature on a proxy.

[4]   Donovan, S., "The SIP INFO Method," RFC 2976, 2000.

[5]   Rosenberg, J., and H. Schulzrinne, "Reliability of Provisional Responses," IETF Internet-Draft, Work in Progress.

[6]   Hoffman, P., L. Masinter, and J. Zawinski, "The mailto URL Scheme," RFC 2368, 1998.

[7]   Freed, N., and N. Borenstein, "Multipurpose Internet Mail Extensions (MIME) Part One: Format of Internet Message Bodies," RFC 2045, 1996.

5

SIP Response Messages

This chapter covers the types of SIP response messages. A SIP response is a message generated by a UAS or a SIP server to reply to a request generated by a UAC. A response may contain additional headers containing information needed by the UAC. Or, it may be a simple acknowledgement to prevent retransmissions of the request by the UAC. Many responses direct the UAC to take specific additional steps. The responses are discussed in terms of structure and classes. Then, each request type is discussed and examined in detail.

There are six classes of SIP responses. The first five classes were borrowed from HTTP; the sixth was created for SIP. The classes are shown in Table 5.1.

If a particular SIP response code is not understood by a UAC, it must be interpreted by the class of the response. For example, an unknown 599 Server Unplugged response must be interpreted by a user agent as a 500 Server Failure response.

The reason phrase is for human consumption only—the SIP protocol uses only the response code in determining behavior. Thus, a 200 Call Failed is interpreted the same as 200 OK. The reason phrases listed here are the suggested ones from the standard document [1]. They can be used to convey more information, especially in failure class responses—the phrase is likely to be displayed to the user. Response codes in the range x00–x79 were borrowed from HTTP, perhaps with a slightly different reason

Table 5.1

SIP Response Classes

Class	Description	Action
1xx	Informational: Indicates status of call prior to completion, also called provisional	If first informational response, the client should switch from timer T1 to timer T2 for retransmission
2xx	Success: request has succeeded	If for an INVITE, ACK should be sent; otherwise, stop retransmissions of request
3xx	Redirection: server has returned possible locations	The client should retry request at another server
4xx	Client error: the request has failed due to an error by the client	The client may retry the request if reformulated according to response
5xx	Server failure: the request has failed due to an error by the server	The request may be retried at another server
6xx	Global failure: the request has failed	The request should not be tried again at this or other servers

phrase[1]. New response codes created for SIP begin at x80 to avoid conflicts with future HTTP response codes.

5.1 Informational

The informational class of responses 1xx are used to indicate call progress. Informational responses are end-to-end responses and may contain message bodies. The exception to this is the 100 Trying response, which is only a hop-by-hop response. Any number of informational responses can be sent by a UAS prior to a final response (2xx 6xx) being sent. The first informational response received by the UAC confirms receipt of the INVITE, and stops retransmission of the INVITE, as described in Section 3.5. For this reason, the standard recommends that a server return a 1xx response if it may take longer than 200ms before it sends a final response. This minimizes INVITE

---

1. Not all HTTP response codes are supported in SIP. For example, HTTP supports a number of success class responses—only one (200 OK) makes sense in SIP. Only the response codes described in RFC 2543 and supporting RFCs are supported in SIP.

retransmissions in the network. Further informational responses have no effect on INVITE retransmissions. A stateful proxy receiving a retransmission of an INVITE will resend the last provisional response sent to date. Informational responses are optional—a UAS can send a final response without first sending an informational response. While final responses to an INVITE receive an ACK to confirm receipt, provisional responses are not acknowledged, except using the PRACK method described in Section 4.1.8. For non-INVITE requests, a 1xx response changes the UAC retransmission timer from T1 to T2.

5.1.1   100 Trying

This special case response is only a hop-by-hop request. It is never forwarded and may not contain a message body. A forking proxy must send a 100 Trying response, since the extended search being performed may take a significant amount of time. This response can be generated by either a proxy server or a user agent. It only indicates that some kind of action is being taken to process the call—it does not indicate that the user has been located.

5.1.2   180 Ringing

This response is used to indicate that the INVITE has been received by the user agent and that alerting is taken place. This response is important in interworking with telephony protocols, and it is typically mapped to messages such as an ISDN Progress or ISUP Address Complete Message (ACM) [2]. For this reason, it might be sent reliably and acknowledged using the PRACK method. When the user agent answers immediately, a 200 OK is sent without a 180 Ringing; this scenario is called the "fast answer" case in telephony.

A message body in this response could be used to carry QoS or security information, or to convey ring tone or animations from the UAS to the UAC.

5.1.3   181 Call Is Being Forwarded

This response is used to indicate that the call has been handed off to another end-point. This response is sent when this information may be of use to the caller. Also, because a forwarding operation may take longer for the call to be answered, this response gives a status for the caller.

5.1.4   182 Call Queued

This response is used to indicate that the INVITE has been received, and will be processed in a queue. The reason phrase can be used to indicate the estimated wait time or the number of callers in line, as shown in Figure 5.1.

A message body in this response can be used to carry music on hold or other media.

5.1.5   183 Session Progress

The 183 Session Progress [3] response indicates that information about the progress of the session (call state) is present in the message body media information. A typical use of this response is to allow a UAC to hear ring tone, busy tone, or a recorded announcement in calls through a gateway into the PSTN. This is because call progress information is carried in the media stream in the PSTN. A one-way media connection or trunk is established from the calling party's telephone switch to the called party's telephone switch in the PSTN prior to the call being answered. In SIP, the media session is established after the call is answered—after a 200 OK and ACK have been exchanged between the UAC and UAS. If a gateway used a 180 Ringing response instead, no media path would be established between the UAC and the gateway, and the

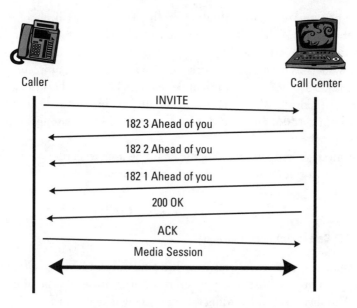

Figure 5.1 Call queuing example with call processing center.

caller would never hear ring tone, busy tone, or a recorded announcement (e.g., "The number you have dialed has changed, the new number is ...") since these are all heard in the media path prior to the call being answered. Figure 5.2 shows an example where a SIP caller does not hear a recorded announcement coming from the PSTN. Figure 5.3 shows the use of the 183 Session Progress allowing an early media session to be established prior to the call being answered. The PSTN interworking scenarios in Chapter 9 show this in detail.

5.2 Success 200 OK

There is only one success class response defined currently in SIP: 200 OK. The success class response has two uses in SIP. When used to accept a session

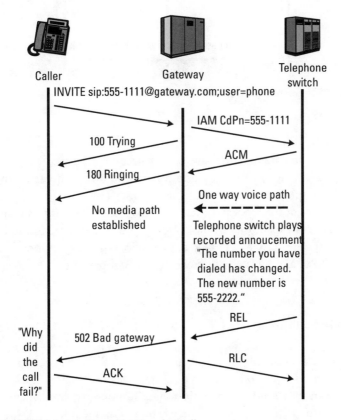

Figure 5.2 PSTN interworking without early media.

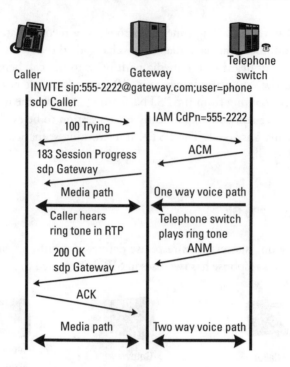

Figure 5.3 PSTN interworking with early media.

invitation, it will contain a message body containing the media properties of the UAS (called party). When used in response to other requests, it indicates successful completion or receipt of the request. The response stops further retransmissions of the request. In response to an OPTIONS, the message body may contain the capabilities of the server. A message body may also be present in a response to a REGISTER request. For 200 OK responses to BYE, CANCEL, INFO, and PRACK, a message body is not permitted.

5.3 Redirection

Redirection class responses are generally sent by a SIP server acting as a redirect server in response to an INVITE, as described in Section 3.3.2. A UAS, however, can also send a redirection class response to implement certain types of call forwarding features. There is no requirement that a UAC receiving a redirection response must retry the request to the specified address. The

UAC could be configured to automatically generate a new INVITE upon receipt of a redirection class response without requiring user intervention.

To prevent looping, the server must not return any addresses contained in the request Via header, and the client must check the address returned in the Contact header against all other addresses tried in an earlier call attempt. Note that this type of transaction looping is different from request looping.

5.3.1   300 Multiple Choices

This redirection response contains multiple Contact headers, which indicate that the location service has returned multiple possible locations for the SIP URL in the Request-URI. The order of the Contact headers is assumed to be significant. That is, they should be tried in the order in which they were listed in the response.

5.3.2   301 Moved Permanently

This redirection response contains a Contact header with the new permanent URL of the called party. The address can be saved and used in future INVITE requests.

5.3.3   302 Moved Temporarily

This redirection response contains a URL that is currently valid but that is not permanent. As a result, the Contact header should not be cached across calls unless an Expires header is present, in which case the location is valid for the duration of the time specified.

5.3.4   305 Use Proxy

This redirection response contains a URL that points to a proxy server who has authoritative information about the calling party. The caller should resend the request to the proxy for forwarding. This response could be sent by a UAS that is using a proxy for incoming call screening. Because the proxy makes the decisions for the UAS on acceptance of the call, the UAS will only respond to INVITE requests that come from the screening proxy. Any INVITE request received directly would automatically receive this response without user intervention.

5.3.5   380 Alternative Service

This response returns a URL that indicates the type of service that the called party would like. An example might be a redirect to a voicemail server.

5.4   Client Error

This class of response is used by a server or UAS to indicate that the request cannot be fulfilled as it was submitted. The specific client error response or the presence of certain headers should indicate to the UAC the nature of the error and how the request can be reformulated. The UAC should not resubmit the request without modifying the request based on the response. The same request, however, can be tried in other locations. A forking proxy receipt of a 4xx response does not terminate the search. Typically, client error responses will require user intervention before a new request can be generated.

5.4.1   400 Bad Request

This response indicates that the request was not understood by the server. An example might be a request that is missing required headers such as To, From, Call-ID, or CSeq. This response is also used if a UAS receives multiple INVITE requests (not retransmissions) for the same Call-ID.

5.4.2   401 Unauthorized

This response indicates that the request requires the user to perform authentication. This response is generally sent by a user agent, since the 407 Proxy Authentication Required (Section 5.4.8) is sent by a proxy that requires authentication. The exception is a registrar server, which sends a 401 Unauthorized response to a REGISTER message that does not contain the proper credentials. An example of this response is:

```
SIP/2.0 401 Unathorized
Via: SIP/2.0/UDP proxy.globe.org:5060;branch=2311ff5d.1
Via: SIP/2.0/UDP 173.23.43.1:5060
To: <sip:printer@maps-r-us.com>;tag=19424103
From: Copernicus <sip:copernicus@globe.org>
Call-ID: 123456787@173.23.43.1
CSeq: 1 INVITE
WWW-Authenticate: Digest realm="Global Phone Company",
  domain = "globe.org" , nonce="8eff88df84f1cec4341ae6e5a359",
```

```
opaque="", stale=FALSE, algorithm=MD5
Content-Length: 0
```

The presence of the WWW-Authenticate header is required to give the calling user agent a chance to respond with the correct credentials. A typical authentication exchange using SIP digest is shown in Figure 3.7. Note that the follow-up INVITE request may use the same Call-ID as the original request. In fact, some authentication implementations may fail if the Call-ID is changed from the initial request to the retried request with the proper credentials.

5.4.3 402 Payment Required

This response is a placeholder for future definition in the SIP protocol. It could be used to negotiate call completion charges.

5.4.4 403 Forbidden

This response is used to deny a request without giving the caller any recourse. It is sent when the server has understood the request, found the request to be correctly formulated, but will not service the request. This response is not used when authorization is required.

5.4.5 404 Not Found

This response indicates that the user identified by the SIP URL in the Request-URI can not be located by the server, or that the user is not currently signed on with the user agent.

5.4.6 405 Method Not Allowed

This response indicates that the server or user agent has received and understood a request but is not willing to fulfill the request. An example might be a REGISTER request sent to a user agent, or an INFO request sent to a proxy server. An Allow header must be present to inform the UAC what methods are acceptable. This is different from the case of an unknown method, in which a 500 Server Error response is returned. Note that a proxy will forward request types it does not understand.

5.4.7   406 Not Acceptable

This response indicates that the request can not be processed due to a requirement in the request message. The Accept header in the request did not contain any options supported by the UAS.

5.4.8   407 Proxy Authentication Required

This request sent by a proxy indicates that the UAC must first authenticate itself with the proxy before the request can be processed. The response should contain information about the type of credentials required by the proxy in a Proxy-Authenticate header. The request can be resubmitted with the proper credentials in a Proxy-Authorization header. Unlike in HTTP, this response may not be used by a proxy to authenticate another proxy.

```
SIP/2.0 407 Proxy Authorization Required
Via: SIP/2.0/UDP discrete.sampling.org:5060
From: Shannon <sip:shannon@sampling.org>
To: Schockley <sip:shockley@transistor.com>;tag=1
Call-ID: adf8gasdd7fld@discrete.sampling.org
CSeq: 1 INVITE
Proxy-Authenticate: Digest realm="SIP",
 domain=sampling.org, nonce=
       "9c8e88df84df1cec4341ae6cbe5a359",
 opaque="", stale="FALSE", algorithm="MD5"
Content-Length: 0
```

5.4.9   408 Request Timeout

This response is sent when an Expires header is present in an INVITE request, and the specified time period has passed. This response could be sent by a forking proxy or a user agent. The request can be retried at any time by the UAC, perhaps with a longer time period in the Expires header or no Expires header at all.

5.4.10   409 Conflict

This response indicates that the request cannot be processed due to a conflict in the request. This response is used by registrars to reject a registration with a conflicting action parameter.

5.4.11 410 Gone

This response is similar to the 404 Not Found response but contains the hint that the requested user will not be available at this location in the future. This response could be used by a service provider when a user cancels their service.

5.4.12 411 Length Required

This response can be used by a proxy to reject a request containing a message body but no Content-Length header. A proxy that takes a UDP request and forwards it as a TCP request could generate this response, since the use of Content-Length is more critical in TCP requests.

5.4.13 413 Request Entity Too Large

This response can be used by a proxy to reject a request that has a message body that is too large. A proxy suffering congestion could temporarily generate this response to save processing long requests.

5.4.14 414 Request-URI Too Long

This response indicates that the Request-URI in the request was too long and cannot be processed correctly. There is no maximum length defined for a Request-URI in the SIP standard document.

5.4.15 415 Unsupported Media Type

This response sent by a user agent indicates that the media type contained in the INVITE request is not supported. For example, a request for a video conference to a PSTN gateway that only handles telephone calls will result in this response. The response should contain headers to help the UAC reformulate the request.

5.4.16 420 Bad Extension

This response indicates that the extension specified in the Require header is not supported by the proxy or user agent. The response should contain a Supported [3] header listing the extensions that are supported. The UAC could resubmit the same request without the extension in the Require header or submit the request to another proxy or user agent.

5.4.17 421 Extension Required

This response [4] indicates that a server requires an extension to process the request that was not present in a Supported header in the request. The required extension should be listed in a Required header in the response. The client should retry the request adding the extension to a Supported header, or try the request at a different server that may not require the extension.

5.4.18 480 Temporarily Unavailable

This response indicates that the request has reached the correct destination, but the called party is not available for some reason. The reason phrase should be modified for this response to give the caller a better understanding of the situation. The response should contain a Retry-After header indicating when the request may be able to be fulfilled. For example, this response could be sent when a telephone has its ringer turned off, or a "do not disturb" button has been pressed. This response can also be sent by a redirect server.

5.4.19 481 Call Leg/Transaction Does Not Exist

This response indicates that a response referencing an existing call or transaction has been received for which the server has no records or state information.

5.4.20 482 Loop Detected

This response indicates that the request has been looped and has been routed back to a proxy that previously forwarded the request. This loop detection is done in a stateless way in SIP using the Via headers. Each server that forwards a request adds a Via header with its address to the top of the request. A branch parameter is added to the Via header, which is a hash function of the Request-URI, and the To, From, Call-ID, and CSeq number. A second part is added to the branch parameter if the request is being forked.

Upon receipt of a request, a proxy will search the Via headers for one with its own address. If one matches, and the branch computed from the current request matches the branch in the Via header, the request is discarded and this response is returned to the UAC. The reason the branch parameter must be checked is to allow a request to be routed back to a proxy, provided that the Request-URI has changed. This could happen with a call forwarding feature. In this case, the Via headers would differ by having different branch parameters.

5.4.21 483 Too Many Hops

This response indicates that the request has been forwarded the maximum number of times as set by the Max-Forwards header in the request. This is indicated by the receipt of a Max-Forwards: 0 header in a request. In the following example, the UAC included a Max-Forwards: 4 header in the REGISTER request. A proxy receiving this request five hops later generates a 483 response:

```
REGISTER sip:registrar.timbuktu.tu SIP/2.0
Via: SIP/2.0/UDP 201.202.203.204:5060;branch=45347.1
Via: SIP/2.0/UDP 198.20.2.4:6128;branch=917a4d4.1
Via: SIP/2.0/UDP 18.56.3.1:5060;branch=7154.1
Via: SIP/2.0/TCP 101.102.103.104:5060;branch=a5ff4d3.1
Via: SIP/2.0/UDP 168.4.3.1:5060
To: sip:explorer@geographic.org
From: sip:explorer@geographic.org
Call-ID: 67483010384@168.4.3.1
CSeq: 1 REGISTER
Max-Forwards: 0
Contact: sip:explorer@national.geographic.org
Content-Length: 0

SIP/2.0 483 Too Many Hops
Via: SIP/2.0/UDP 201.202.203.204:5060;branch=45347.1
Via: SIP/2.0/UDP 198.20.2.4:6128;branch=917a4d4.1
Via: SIP/2.0/UDP 18.56.3.1:5060;branch=7154.1
Via: SIP/2.0/TCP 101.102.103.104:5060;branch=a5ff4d3.1
Via: SIP/2.0/UDP 168.4.3.1:5060
To: <sip:explorer@geographic.org>;tag=a5642
From: sip:explorer@geographic.org
Call-ID: 67483010384@168.4.3.1
CSeq: 1 REGISTER
Content-Length: 0
```

5.4.22 484 Address Incomplete

This response indicates that the Request-URI address is not complete. This could be used in an overlap dialing scenario in PSTN interworking where digits are collected and sent until the complete telephone number is assembled by a gateway and routed [4]. Note that the follow-up INVITE requests may use the same Call-ID as the original request. An example of overlap dialing is shown in Figure 5.4.

5.4.23 485 Ambiguous

This request indicates that the Request-URI was ambiguous and must be clarified in order to be processed. This occurs if the username matches a number of registrations. If the possible matching choices are returned in Contact headers, then this response is similar to the 300 Multiple Choices response. They are slightly different, however, since the 3xx response returns equivalent choices for the same user, but the 4xx response returns alternatives that can be different users. The 3xx response can be

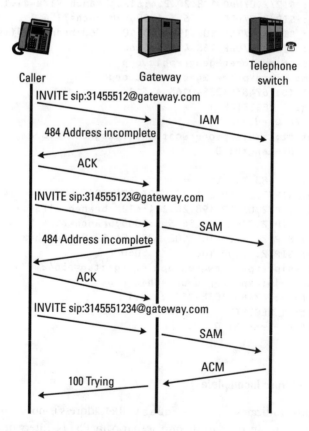

Figure 5.4 Overlap dialing to the PSTN with SIP.

processed without human intervention, but this `4xx` response requires a choice by the caller, which is why it is classified as a client error class response. A server configured to return this response must take user registration privacy into consideration, otherwise a vague or general `Request-URI` could be used by a rogue user agent to try to discover SIP URLs of registered users.

5.4.24　486 Busy Here

This response is used to indicate that the user agent cannot accept the call at this location. This is different, however, from the `600 Busy Everywhere` response, which indicates that the request should not be tried elsewhere. In general, a `486 Busy Here` is sent by a UAS unless it knows definitively that the user cannot be contacted. This response is equivalent to the busy tone in the PSTN.

5.4.25　487 Request Canceled

This response can be sent by a user agent that has received a `CANCEL` request for a pending `INVITE` request. A `200 OK` is sent to acknowledge the `CANCEL`, and a `487` is sent in response to the `INVITE`.

5.4.26　488 Not Acceptable Here

This response indicates that some aspect of the proposed session is not acceptable and may contain a `Warning` header indicating the exact reason. This response has a similar meaning to `606 Not Acceptable`, but only applies to one location and may not be true globally as the `606` response indicates.

5.5　Server Error

This class of responses is used to indicate that the request cannot be processed because of an error with the server. The response may contain a `Retry-After` header if the server anticipates being available within a specific time period. The request can be tried at other locations because there are no errors indicated in the request.

5.5.1   500 Server Internal Error

This server error class response indicates that the server has experienced some kind of error that is preventing it from processing the request. The reason phrase can be used to identify the type of failure. The client can retry the request again at this server after several seconds.

5.5.2   501 Not Implemented

This response indicates that the server is unable to process the request because it is not supported. This response can be used to decline a request containing an unknown method. A proxy, however, will forward a request containing an unknown request method. Thus, a proxy will forward an unknown SELF-DESTRUCT request, assuming that the UAS will generate this response if the method is not known.

5.5.3   502 Bad Gateway

This response is sent by a proxy that is acting as a gateway to another network, and indicates that some problem in the other network is preventing the request from being processed.

5.5.4   503 Service Unavailable

This response indicates that the requested service is temporarily unavailable. The request can be retried after a few seconds, or after the expiration of the Retry-After header. Instead of generating this response, a loaded server may just refuse the connection.

5.5.5   504 Gateway Timeout

This response indicates that the request failed due to a timeout encountered in the other network to which that the gateway connects. It is a server error class response because the call is failing due to a failure of the server in accessing resources outside the SIP network.

5.5.6   505 Version Not Supported

This response indicates that the request has been refused by the server because of the SIP version number of the request. The detailed semantics of this response have not yet been defined because there is only one version of SIP (version 2.0) currently implemented. When additional version numbers are implemented in the future, the mechanisms for dealing with multiple protocol versions will need to be detailed.

5.6 Global Error

This response class indicates that the server knows that the request will fail wherever it is tried. As a result the request should not be sent to other locations. Only a server that has definitive knowledge of the user identified by the Request-URI in every possible instance should send a global error class response. Otherwise, a client error class response should be sent. A Retry-After header can be used to indicate when the request might be successful.

5.6.1 600 Busy Everywhere

This response is the definitive version of the 486 Busy Here client error response. If there is a possibility that the call to the specified Request-URI could be answered in other locations, this response should not be sent.

5.6.2 603 Decline

This response has the same effect as the 600 Busy Everywhere but does not give away any information about the call state of the server. This response could indicate the called party is busy, or simply does not want to accept the call.

5.6.3 604 Does Not Exist Anywhere

This response is similar to the 404 Not Found response but indicates that the user in the Request-URI cannot be found anywhere. This response should only be sent by a server that has access to all information about the user.

5.6.4 606 Not Acceptable

This response can be used to implement some session negotiation capability in SIP. This response indicates that some aspect of the desired session is not acceptable to the UAS, and as a result, the session cannot be established. The response may contain a Warning header with a numerical code describing exactly what was not acceptable. The request can be retried with a different media session information. An example of simple negotiation with SIP is shown in Figure 5.5. If more complicated negotiation capability is required, another protocol should be used.

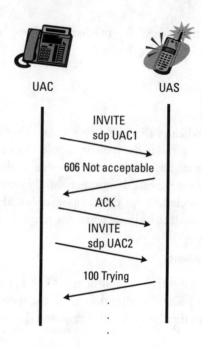

Figure 5.5 Session negotiation with SIP.

References

[1]   Handley, M., et al., "SIP: Session Initiation Protocol," RFC 2543, 1999.

[2]   Anttalainen, T., *Introduction to Telecommunications Network Engineering,* Artech House: Norwood, MA, 1999.

[3]   Donovan, S., et al., "SIP 183 Session Progress Message," IETF Internet-Draft, Work in Progress.

[4]   Rosenberg, J., and H. Schulzrinne, "The SIP Supported Header," IETF Internet-Draft, Work in Progress.

[5]   Camarillo, G., and A. Roach, "Best Current Practice for ISUP to SIP Mapping," IETF Internet-Draft, Work in Progress.

6

SIP Headers

This chapter describes the headers present in SIP messages. There are four types of SIP headers: general, request, response, and entity. They are described using this grouping in the following sections. Except as noted, headers are defined in the SIP specification [1].

SIP headers in most cases follow the same rules as HTTP headers [2]. Headers are defined as `header:field` where `header` is the case-insensitive token used to represent the header, and `field` is the case-insensitive set of tokens that contain the information. Except when otherwise noted, their order in a message is not important. Header fields can continue over multiple lines as long as the line begins with at least one space or horizontal tab character. Unrecognized headers are ignored by proxies. Many common SIP headers have a compact form, where the header name is denoted by a single lower-case character. These headers are shown in Table 6.1

Headers can be either end-to-end or hop-by-hop. Hop-by-hop headers are the only ones that a proxy may insert, or with a few exceptions, modify. A proxy should never change the header order. Because SIP typically involves end-to-end control, most headers are end-to-end. The hop-by-hop headers that may be inserted by a proxy are shown in Table 6.2.

6.1 General Headers

The set of general headers includes all of the required headers in a SIP message. General headers can be present in both requests and responses. These

Table 6.1
Compact Forms of SIP Headers

Header	Compact Form
Call-ID	i
Contact	m
Content-Encoding	e
Content-Length	l
Content-Type	c
From	f
Subject	s
To	t
Via	v

Table 6.2
Hop-by-Hop Headers that May Be Inserted by Proxies

Hop-by-hop headers
Hide
Organization
Proxy-Authenticate
Proxy-Authorization
Proxy-Require
Record-Route
Route

headers are created by user agents and cannot be modified by proxies, with a few exceptions.

6.1.1   Call-ID

The Call-ID header is mandatory in all SIP requests and responses. It is part of the call leg used to uniquely identify a call between two user agents. A

Call-ID must be unique across calls, except in the case of a Call-ID in registration requests. All registrations for a user agent should use the same Call-ID. A Call-ID is always created by a user agent and is never modified by a server.

The Call-ID is usually made up of a local-id, which should be a cryptographically random identifier, the @ symbol, and a host name or IP address. Because a user agent can ensure that its local-id is unique within its domain, the addition of the globally unique host name makes the Call-ID globally unique. Some security is provided by the randomness of the Call-ID, because this prevents a third party from *guessing* a Call-ID and presenting false requests. The compact form of the Call-ID header is i.

Examples of Call-ID are shown in Table 6.3.

6.1.2 Contact

The Contact header is used to convey a URL that identifies the resource requested or the request originator, depending on whether it is present in a request or response. Once a Contact header has been received, that URL can be cached and used for routing future requests. For example, a Contact header in a 200 OK response to an INVITE can allow the acknowledgment ACK message and all future requests during this call to bypass proxies and go directly to the called party. However, the presence of Record-Route headers in an earlier request or default proxy routing configuration in a user agent may override that behavior. When a Contact URL is used in a Request-URI, all URL parameters are allowed with the exception of the method parameter, which is ignored.

Table 6.3
Examples of Call-ID Headers

Header	Meaning
Call-ID: 34a5d553192cc35@15.34.3.1	A hexadecimal UUID is used, along with a host IPv4 address
Call-ID: 2000-JUL-07-23-12@digitalari.com	A local-id made up of a time stamp along with a domain name
i: 35866383092031257@port34.carrier.com	The compact form is used, with a local-id as a random decimal number

Contact headers must be present in INVITE requests and 200 OK responses to invitations. If the user agent is behind a firewall, the Contact address will be the firewall proxy address. Otherwise, the use of the user agent's URL will result in the call failing because of the firewall blocking any direct routed SIP requests. Contact headers may also be present in 1xx, 2xx, 3xx, and 485 responses. Only in a REGISTER request, a special Contact:*, along with an Expires: 0, header is used to remove all existing registrations. Examples of Contact headers in registrations are shown in Table 4.3. Otherwise, wild carding is not allowed. A Contact header may contain a display name that can be in quotes. If a display name is present, the URL will be enclosed in <>. If any header parameters are present, the URL will also be enclosed in <>, with the header parameters outside the <>, even if no display name is present.

There are three additional parameters defined for use in Contact headers: q, action, and expires. They are placed at the end of the URL or URI and separated by semicolons.

The qvalue parameter is used to indicate relative preference, which is represented by a decimal number in the range 0 to 1. The qvalue is not a probability, and there is no requirement that the qvalues for a given list of Contacts add up to 1. The action parameter is only used in registration Contact headers, and is used to specify proxy or redirect operation by the server. The expires parameter indicates how long the URL is valid and is also only used in registrations. The parameter either contains an integer number of seconds or a date in SIP form (see Section 6.1.4). Examples are shown in Table 6.4.

6.1.3   CSeq

The command sequence CSeq header is a required header in every request. The CSeq header contains a decimal number that increases for each request. Usually, it increases by 1 for each new request, with the exception of CANCEL and ACK requests, which use the CSeq number of the INVITE request to which it refers.

The CSeq count is used by UASs to determine out-of-sequence requests or to differentiate between a new request (different CSeq) or a retransmission (same CSeq). The CSeq header is used by UACs to match a response to the request it references. For example, a UAC that sends an INVITE request then a CANCEL request can tell by the method in the CSeq of a 200 OK response if it is a response to the invitation or cancellation request. Examples are shown in Table 6.5.

Table 6.4

Examples of Contact Headers

Header	Meaning
Contact: sip:bell@telephone.com	A single SIP URL without a display name.
Contact: Lentz <h.lentz@petersburg.edu>	A display name with the URL is enclosed in < > ; the display name is treated as a token and ignored.
Contact: M. Faraday <faraday@effect.org>, "Faraday" <mailto:faraday@pop.effect.org>	Two URLs are listed, the second being a non-SIP URL with a display name enclosed in quotes.
m: <morse@telegraph.org; transport=tcp>; expires= "Fri, 13, Oct 1998 12:00:00 GMT"	The compact form of the header is used for a single URL. The URL contains a URL parameter contained within the < >. An expires header parameter uses a SIP date enclosed in the quotes.

Table 6.5

CSeq Header Examples

Header	Meaning
CSeq: 1 INVITE	The command sequence number has been initialized to 1 for this initial INVITE
CSeq: 432 INFO	The command sequence number is set to 432 for this INFO request
CSeq: 6787 INVITE	If this was the first request by the user agent for this *Call-ID,* then either the *CSeq* was initialized to 6787, or the previous request generated for this *Call-ID* (either an INVITE or other request) would have had a CSeq of *6786* or lower

Each user agent maintains its own command sequence number space. For example, consider the case where user agent 1 establishes a session to user agent 2 and initializes its CSeq count to 1. When user agent 2 initiates a request (such as a INVITE or INFO, or even BYE) it will initialize its own CSeq space, totally independent of the CSeq count used by user agent 1. The examples of Chapter 9 show this behavior of CSeq.

6.1.4 Date

The Date header is used to convey the date when a request or response is sent. The format of a SIP date is based on HTTP dates, but allows only the preferred Internet date standard referenced by RFC 1123 [3]. To keep user agent date and time logic simple, SIP only supports the use of the GMT time zone. This allows time entries that are stored in date form rather than second count to be easily converted into delta seconds without requiring knowledge of time zone offsets. A Date example is shown below:

```
Date: Fri, 13 Oct 1998 23:29:00 GMT
```

6.1.5 Encryption

The Encryption header is used to specify the portion of a SIP message that has been encrypted. All information after the header is assumed to have been encrypted. Encryption provides a level of privacy for end users who wish to keep media information private from third parties who could intercept SIP INVITE or 200 OK messages. The encryption is done using the public key of the recipient of the request (as identified by the To header). The public key is carried in the Response-Key header, but the private key is transmitted using some non-SIP method. Only headers that are not used by proxies in routing requests may be encrypted. Because proxies can have logic to use almost any header field to determine routing, there is no way to be sure that an encrypted request will be processed by a proxy in the same way as an unencrypted one.

An INVITE request with an encrypted message body is shown below:

```
INVITE sip:schottky@diode.org SIP/2.0
Via: sip/2.0/UDP room203.lab.bell.com:5060
From: <sip:brattain@lab.bell.com>
To: <sip:schottky@diode.org>
Call-ID: 3a34-d654-21b6-49f0@lab.bell.com
Content-Length: 175
Encryption: MD5

84d7c249cfa12febedb30da41f9c12eb3db7f039e2f38a4716e4
012040123be09109c5871cc2cd0848b99bc8399e9d33bf8e1d9bf
662b190bf6199c01034e98f54eb0d989d84c570d9a47980e22357
ed23d09ba452e34
```

6.1.6 From

The From header is a required header that indicates the originator of the request. It is one of two addresses used to identify the call leg. The From header contains a URL, but it may not contain the transport, maddr, or ttl URL parameters. A From header may contain a tag, used to identify a particular call. If it is possible that two end-points may have multiple calls between them with identical Call-IDs, then the initiator of the session must include a tag in the From header. A From header may contain a display name, in which case the URL is enclosed in <>. If there is both a URL parameter and a tag, then the URL including any parameters must be enclosed in <>. Examples are shown in Table 6.6.

6.1.7 Organization

The Organization header is used to indicate the organization to which the originator of the message belongs. It can also be inserted by proxies as a message is passed from one organization to another. Like all SIP headers, it can be used by proxies for making routing decisions and by user agents for making call screening decisions. An example is below:

```
Organization: WorldCom
```

Table 6.6
Examples of From Header

Header	Meaning
From: sip:armstrong@hetrodyne.com	A single SIP URL without a display name.
From: Thomas Edison <sip:edison@electric.com>	A display name is used, so the URL is enclosed in < > ; the display name is treated as a token and ignored
f: "James Bardeen" <555.1313@telephone.com;user=phone>	Using the compact form of the header, a display name in quotes along with a SIP URL with a parameter inside the < >.
From:<911@emergency.com; user=phone> ;tag=d632a2	Both a URL parameter and tag are used, so URL is enclosed in < >.

6.1.8   Retry-After

The Retry-After header is used to indicate when a resource or service may be available again. In 503 Service Unavailable responses, it indicates when the server will be available. In 404 Not Found, 600 Busy Everywhere, and 603 Decline responses, it indicates when the called user agent may be available again.

The header can also be included by proxy and redirect servers in responses if a recent registration was removed with a Retry-After header indicating when the user may sign on again. The contents of the header can be either an integer number of seconds or a SIP date. An optional comment enclosed in () can be included to provide more information. A duration parameter can be used to indicate how long the resource will be available after the time specified. Examples of this header are shown in Table 6.7.

6.1.9   Subject

The optional Subject header is used to indicate the subject of the media session. It can be used by user agents to do simple call screening. The contents of the header can also be displayed during alerting to aid the user in deciding whether to accept the call. The compact form of this header is s. Some examples are:

```
Subject: More good info about SIP
s: Are you awake, yet??
```

Table 6.7
Examples of Retry-After Header

Header	Meaning
Retry-After: 3600	Request can be retried again in 1 hour.
Retry-After: Sat, 21 May 2000 08:00:00 GMT	Request can be retried after the date listed.
Retry-After: Fri, 1 Oct 2000 18:05:00 GMT (I'm at lunch.)	Header contains comments.
Retry-After: Mon, 29 Feb 2000 13:30:00 GMT; duration=1800	Request can be retried after the specified date for 30 minutes.

6.1.10 Supported

The Supported header [4] is used to list one or more options implemented by a user agent or server. It is typically included in responses to OPTIONS requests. If no options are implemented, the header is not included. If a UAC lists an option in a Supported header, proxies or UASs may use the option during the call. If the option must be used or supported, the Require header is used instead. An example of the header is:

    Supported: rel100

6.1.11 Timestamp

The Timestamp header is used by a UAC to mark the exact time a request was generated in some numerical time format. A UAS must echo the header in the response to the request and may add another numerical time entry indicating the amount of delay. Unlike the Date header, the time format is not specified. The most accurate time format should be used, including a decimal point. Examples are shown in Table 6.8.

6.1.12 To

The To header is a required header in every SIP message used to indicate the recipient of the request. Any responses generated by a user agent will contain this header with the addition of a tag if more than one Via header is present, as described in Section 4.3. Any response generated by a proxy must have a tag added to the To header. If a tag has been added to the header in a 200 OK response, it is used throughout the call and incorporated into the call leg. The To header is never used for routing—the Request-URI is used for

Table 6.8
Examples of Timestamp Header

Header	Meaning
Timestamp: 235.15	Client has stamped a start time for the request
Timestamp: 235.15 .95	This header from the response has the delay time added by the server

this purpose. An optional display name can be present in the header, in which case the SIP URL is enclosed in <>. If the URL contains any parameters or username parameters, the URL must be enclosed in <> even if no display name is present. The compact form of the header is *t*. Examples are shown in Table 6.9.

6.1.13 User Agent

The User-Agent header is used to convey information about the user agent originating the request. Based on the HTTP header of the same name [2], it can contain manufacturer information, software version, or comments. The field may contain multiple tokens, with the ordering assumed to be from most general to most specific. This information can be used for logging or for generating a specific response for a specific user agent. Examples include:

```
User-Agent: Acme SIP Phone v2.2
User-Agent: IP Carrier Gateway Av6.4
```

6.1.14 Via

The required Via header is used to record the SIP route taken by a request and is used to route a response back to the originator. A user agent generating a request records its own address in a Via header in the request. While the ordering of most SIP headers is not significant, the Via headers order is

Table 6.9
Examples of To Header

Header	Meaning
To: sip:babage@engine.org;tag=2443a8f7	A single SIP URL with a tag and without a display name
To: Thomas Edison <sip:edison@electric.com>	A display name is used, so the URL is enclosed in < >; the display name is treated as a token and ignored
t: "Jim B." <brattain@bell.org>	A display name in quotes along with a SIP URL enclosed within < >.
To: <+1-314-555-1212@carrier.com ;user=phone>;tag=8f7f7ad6675	Both a URL parameter and tag are used, so URL is enclosed in < >.

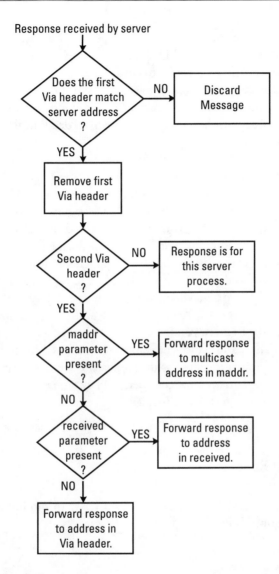

Figure 6.1 Via forwarding decision tree.

significant because it is used to route responses. A proxy forwarding the request adds a Via header containing its own address to the *top* of the list of Via headers. A proxy adding a Via header always includes a branch tag containing a cryptographic hash of the To, From, Call-ID, and Request-URI headers. A proxy or user agent generating a response to a request copies all the Via headers from the request *in order* into the response, then sends the response to the address specified in the top Via header. A proxy receiving a response checks the top Via header to ensure that it matches its own address. If it does not, the response has been misrouted and should be discarded. The top Via header is then removed, and the response forwarded to the address specified in the next Via header. This is shown in detail in the decision tree in Figure 6.1.

Via headers contain protocol name and version number and transport (SIP/2.0/UDP, SIP/2.0/TCP, etc.) and may contain port numbers, and parameters such as received, hidden, branch, maddr, and ttl. A received tag is added to a Via header if a user agent or proxy receives the request from a different address than that specified in the top Via header. This indicates that a NAT or firewall proxy is in the message path. If present, the received tag is used in response routing. The hidden parameter indicates that the Via header has been encrypted. A branch parameter is added to Via headers by proxies, which is computed as a hash function of the Request-URI, and the To, From, Call-ID and CSeq number. A second part is added to the branch parameter if the request is being forked as shown in Figure 3.4. The maddr and ttl parameters are used for multicast transport and have a similar meaning as the equivalent SIP URL parameters. The header can also contain optional comments. The compact form of the header is v. Examples are given in Table 6.10.

6.2   Request Headers

Request headers are added to a request by a UAC to modify or give additional information about the request.

6.2.1   Accept

The Accept header is defined by HTTP [2] and is used to indicate acceptable message Internet media types [4] in the message body. The header describes media types using the format type/sub-type commonly used in the Internet. If not present, the assumed acceptable message body format is application/sdp. A list of media types can have preferences set using qvalue parameters. The wildcard "*" can be used to specify all sub-types. Examples are given in Table 6.11.

Table 6.10

Examples of Via Header

Header	Meaning
Via: SIP/2.0/UDP 100.101.102.103	IPv4 address using unicast UDP transport and assumed port of 5060.
Via: SIP/2.0/TCP cube451. office.com:60202 (My Temporary Office)	Domain name using TCP transport and port number 60202 with a comment in ( ).
Via: SIP/2.0/UDP 120.121.122.123;branch=56a234f3.1	Proxy added Via header with branch.
v: SIP/2.0/UDP proxy.garage.org ;branch=7c8f3423423a3.3	Compact form with domain name using UDP; third search location of forking proxy
Via: SIP/2.0/TCP 192.168.1.2 ;received=12.4.5.50	IPv4 address is non-globally unique. Request has been forwarded through a NAT which changed the IP address to a globally unique one.
Via:SIP/2.0/UDP host.user.com:4321; ;maddr=224.1.2.3 ;ttl=15	The address is a multicast address specified in maddr with a specified TTL.
Via:SIP/2.0/UDP B4hfdWi43koDrtu6sfgl ;hidden	An encrypted Via header.

Table 6.11

Examples of Accept Header

Header	Meaning
Accept: application/sdp	This is the default assumed even if no *Accept* header is present
Accept: text/*	*Accept* all text encoding schemes
Accept: application /h.245;q=0.1, application/sdp;q=0.9	Use SDP if possible, otherwise, use H.245

6.2.2 Accept-Contact

The Accept-Contact [4] header specifies which URLs the request may be proxied to. Some additional parameters are also defined for Contact headers such as media, duplex, and language. This header is part of the caller preferences extensions to SIP, which have been defined to give some control to the caller in the way a proxy server processes a call.

6.2.3 Accept-Encoding

The `Accept-Encoding` header, defined in HTTP [2], is used to specify acceptable message body encoding schemes. Encoding can be used to ensure a SIP message with a large message body fits inside a single UDP datagram. The use of `qvalue` parameters can set preferences. If none of the listed schemes are acceptable to the UAC, a `406 Not Acceptable` response is returned. If not included, the assumed encoding will be `text/plain`. Examples include:

```
Accept-Encoding: text/plain
Accept-Encoding: gzip
```

6.2.4 Accept-Language

The `Accept-Language` header, defined in HTTP [2], is used to specify preferences of language. The languages specified can be used for reason phrases in responses, informational headers such as `Subject`, or in message bodies. The HTTP definition allows the language tag to be made of a primary tag and an optional sub-tag. This header could also be used by a proxy to route to a human operator in the correct language. The language tags are registered by IANA. The primary tag is an ISO-639 language abbreviation. The use of `qvalues` allows multiple preferences to be specified. Examples are shown in Table 6.12.

Table 6.12
Examples of Accept-Language Header

Header	Meaning
Accept-Language: fr	French is the only acceptable language.
Accept-Language: en, ea	Acceptable languages include both English and Spanish
Accept-Language: ea ;q=0.5 en; q=0.9, fr ;q=0.2	Preferred languages are English, Spanish, and French, in that order

6.2.5   Authorization

The Authorization header is used to carry the credentials of a user agent in a request to a server. It can be sent in reply to a 401 Unauthorized response containing challenge information, or it can be sent first without waiting for the challenge if the form of the challenge is known (e.g., if it has been cached from a previous call). The authentication mechanism for SIP digest is described in Section 3.6. When using pretty good privacy (PGP), the PGP signature is calculated across the nonce, realm, request method, request version, and all header fields in order after the Authorization header. Examples are shown in Table 6.13.

6.2.6   Hide

The Hide header is used by user agents or proxies to request that the next hop proxy encrypts the Via headers to hide message routing path information. The header contains either route or hop depending on the type of service requested. The route option is used to request that all future proxies encrypt the message routing information. This is done by each proxy encrypting the current plain view Via headers, then including the Hide: route header in the forwarded request. The hop option is used to request that only the next proxy should perform the encryption. The proxy receiving the request encrypts all the plain view headers, then removes the Hide header before forwarding the request. As the response to the request is routed back

Table 6.13
Examples of Authorization Header

Header	Meaning
Authorization: Digest username="Cust1", realm="SIP Telephone Company", nonce="9c8e88df84f1cec4341ae6e5a359", opaque="", uri="sip:proxy.sip-com.com", response="e56131d19580cd833064787ecc"	This HTTP digest authorization header contains the credentials of Cust1; the nonce was supplied by the SIP server located at the uri specified. The response contains the encrypted username and password. No opaque string is present.
Authorization: pgp version=5.0; realm="Gateway Password Required"; nonce="76e63aff71"; signature="kdD2+kdflk2adfkijfhFWQncvej"	This PGP authorization header contains the PGP version number, a nonce supplied by the challenger and a ASCII-armored signature

through the same set of proxies, the proxies decrypt each Via header they encrypted and use that information to route the response back to the requestor. The following example would result in an encrypted Via header:

```
Hide: hop
```

6.2.7 In-Reply-To

The In-Reply-To header is used to indicate the Call-ID that this request references or is returning. For example, a missed call could be returned with a new INVITE and the Call-ID from the missed INVITE copied into the In-Reply-To header. This allows the UAS to determine that this is not an unsolicited call, which could be used to override call screening logic, for example. Examples of this header are as follows:

```
In-Reply-To: a8-43-73-ff-43@company.com
In-Reply-To: 12934375@persistance.org,
             12934376@persistance.org
```

6.2.8 Max-Forwards

The Max-Forwards header is used to indicate the maximum number of hops that a SIP request may take. The value of the header is decremented by each proxy that forwards the request. A proxy receiving the header with a value of zero discards the message and sends a 483 Too Many Hops response back to the originator. An example is:

```
Max-Forwards: 10
```

6.2.9 Priority

The Priority header is used by a UAC to set the urgency of a request. Defined values are non-urgent, normal, urgent, and emergency. This header could be used to override screening or by servers in load-shedding mechanisms. Because this header is set by the user agent, it may not be possible for a carrier network to use this field to route emergency traffic, for example. An example is:

```
Priority: emergency
```

6.2.10 Proxy-Authorization

The Proxy-Authorization header is to carry the credentials of a user agent in a request to a server. It can be sent in reply to a 407 Proxy Authentication Required response containing challenge information, or it can be sent first without waiting for the challenge if the form of the challenge is known (e.g., if it has been cached from a previous call). The authentication mechanism for SIP digest is described in Section 3.6. When using PGP, the PGP signature is calculated across the nonce, realm, request method, request version, and all header fields in order after the Proxy-Authorization header. A proxy receiving a request containing a Proxy-Authorization header searches for its own realm. If found, it processes the entry. If the credentials are correct, any remaining entries are kept in the request when it is forwarded to the next proxy. An example of this is in Figure 6.2.

Examples are shown in Table 6.14.

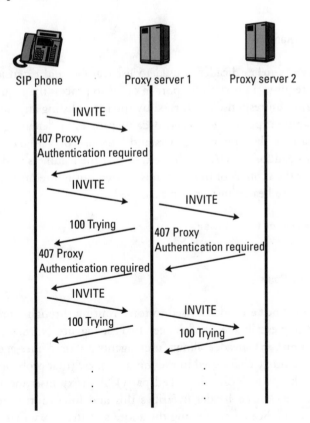

Figure 6.2 Multi-proxy authentication example.

Table 6.14
Examples of Proxy-Authorization Header

Header	Meaning
Proxy-Authorization: Digest username="Customer1", realm="SIP Telephone Company", nonce="9c8e88df84f1cec4341ae6e5a359", opaque="", uri="sip:proxy.sip.com", response="e56131d19580cd833064787ecc"	This digest authorization header contains the credentials of Customer1; the nonce was supplied by the SIP server located at the URI specified; the response contains the encrypted username and password; no opaque string is present.
Proxy-Authorization: pgp version=5.0; realm="Gateway Password Required"; nonce="76e63aff71"; signature="kdD2+kdflk2ajudgHgFWQncvej"	This PGP authorization header contains the PGP version number, a nonce supplied by the challenger, and a ASCII-armored signature

6.2.11 Proxy-Require

The Proxy-Require header is used to list features and extensions that a user agent requires a proxy to support in order to process the request. A 420 Bad Extension response is returned by the proxy listing any unsupported feature in an Unsupported header. Because proxies by default ignore headers and features they do not understand, the use of a Proxy-Require header is needed for the UAC to be certain that the feature is understood by the proxy. If the support of this option is desired but not required, it is listed in a Supported header instead. An example is:

```
Proxy-Require: timer
```

6.2.12 Record-Route

The Record-Route header is used to force routing through a proxy for all subsequent requests in a session between two user agents. Normally, the presence of a Contact header allows user agents to send messages directly bypassing the proxy chain used in the initial request (that probably involved database look-ups to locate the called party). A proxy inserting its address into a Record-Route header overrides this and forces future requests to include a Route header containing the address of the proxy that forces this proxy to be included.

A proxy, such as a firewall proxy, wishing to implement this inserts the header containing its own address, or adds its address to an already present Record-Route header. The URL is constructed from the Request-URI of the request and a maddr parameter containing the URL of the proxy server. The UAS copies the Record-Route header into the 200 OK response to the request. The header is forwarded unchanged by proxies back to the UAC. The UAC then stores the Record-Route proxy list plus a Contact header if present in the 200 OK for use in a Route header in all subsequent requests. Because Record-Route is bi-directional, messages in the reverse direction will also traverse the same set of proxies. Chapter 9 contains an example of the use of the Record-Route and Route headers. Examples are:

```
Record-Route: <sip:me@home.com;maddr=proxy1.carrier.com>,
    <sip:me@home.com;maddr=firewall33.corporation.com>
Record-Route:<sip:visitor54@lobby.hotel.com;maddr=139.23.1.44>
```

6.2.13 Reject-Contact

The Reject-Contact [4] header specifies the URLs to which the request may not be proxied. Some additional parameters are also defined for Contact headers such as media, duplex, and language when used in this header. This header, along with Accept-Contact and Request-Dispositon are part of the SIP caller preferences extensions. An example is:

```
Reject-Contact: sip:admin@boss.com
```

6.2.14 Request-Disposition

The Request-Disposition [4] header can be used to request servers to either proxy or redirect, or initiate serial or parallel (forking) searches. An example is:

```
Request-Disposition: redirect
```

6.2.15 Require

The Require header is used to list features and extensions that a UAC requires a UAS to support in order to process the request. A 420 Bad Extension response is returned by the UAS listing any unsupported features in an Unsupported header. If support or use of a feature is desirable but not required, the Supported header is used instead. An example is:

```
Require: rel100
```

6.2.16 Response-Key

The Response-Key header is used by a client to request that the response to the request should be encrypted with the public key carried in this header. An example is:

```
Response-Key: pgp version="2.6", encoding="ascii",
   key="Ha=tu3f=poe$f1eityu4cnae2g=j6&yudkd[sg8-]
   dkh7uplglyidjk"
```

6.2.17 Route

The Route header is used to force routing for a request through a path specified in the header field. The path was extracted from a Record-Route and/or Contact header received in an earlier request within the same call. Chapter 9 contains an example of the use of the Record-Route and Route headers. Examples of Route headers constructed from the example Record-Route headers in Section 6.2.12 are:

```
Route: <sip:me@home.com;maddr=firewall33.corporation.com>,
   <sip:me@home.com;maddr=proxy1.carrier.com>
Route: <sip:visitor54@lobby.hotel.com;maddr=139.23.1.44>
```

6.2.18 RAck

The RAck header [6] is used within a response to a PRACK request to reliably acknowledge a provisional response that contained a RSeq header. The RAck header echoes the CSeq and the RSeq from the provisional response. The reliable sequence number is incremented for each response sent reliably. An example is:

```
RAck: 8342523 1 INVITE
```

6.2.19 Session-Expires

The Session-Expires header [7] is used to place a time limit on a media session. The header can only be used in INVITE requests and 200 OK responses to INVITEs. The header can only be used if a Supported: timer header was present in the current or previous request or response. A user agent supporting this feature must tear down an existing media session with a

BYE when the session timer started by the Session-Expires header expires. The timer can be extended with a re-INVITE with a new Session-Expires header. Examples are:

```
Session-Expires: 60
Session-Expires: Fri, 1 Apr 2000 11:59:00 GMT
```

6.3 Response Headers

Response headers are added to a response by a UAS or SIP server to give more information than just the response code and reason phrase. They are generally not added to a request.

6.3.1 Proxy-Authenticate

The Proxy-Authenticate header is used in a 407 Proxy Authentication Required authentication challenge by a proxy server to a UAC. It contains the nature of the challenge so that the UAC may formulate credentials in a Proxy-Authorization header in a subsequent request. Examples are shown in Table 6.15.

6.3.2 Server

The Server header is used to convey information about the UAS generating the response. The use and contents of the header are similar to the User-Agent header in Section 6.1.13. An example is:

```
Server: Dotcom Announcement Server B3
```

6.3.3 Unsupported

The Unsupported header is used to indicate features that are not supported by the server. The header is used in a 420 Bad Extension response to a request containing an unsupported feature listed in a Require header. Because multiple features may have been listed in the Require header, the Unsupported header indicates all the unsupported features—the rest can be assumed by the UAC to be supported. An example is:

```
Unsupported: rel100
```

Table 6.15
Examples of Proxy-Authenticate Header

Header	Meaning
Proxy-Authenticate: Digest realm="SIP Telephone Company", domain = "sip-company.com", nonce="9c8e88df84f1cec4341ae6e5a359", opaque="", stale="FALSE", algorithm="MD5"	HTTP digest challenge header
Proxy-Authenticate: pgp version=5.0; realm="Gateway Password Required"; nonce="76e63aff71"; signature="djflk2adfkijfhudgHgFWQncvej"	PGP challenge

6.3.4 Warning

The Warning header is used in a response to provide more specific information than the response code alone can convey. The header contains a three-digit warning code, a warning agent that indicates what server inserted the header, and warning text enclosed in quotes used for display purposes. Warning codes in the 1xx and 2xx range are specific to HTTP [2]. The SIP standard defines 12 new warning codes in the 3xx class. The breakdown of the class is shown in Table 6.16. The complete set of defined warning codes is listed in Table 6.17.

Examples are:

```
Warning: 302 proxy "Incompatible transport protocol"
Warning: 305 room132.hotel.com:5060 "Incompatible media
    type"
```

6.3.5 WWW-Authenticate

The WWW-Authenticate header is used in a 401 Unauthorized authentication challenge by a user agent or registrar server to a UAC. It contains the nature of the challenge so that the UAC may formulate credentials in a Proxy-Authorization header in a subsequent request. SIP supports HTTP basic and digest authentication mechanisms, as well as PGP [8]. Examples are shown in Table 6.18.

Table 6.16
SIP Warning Code Breakdown

Warning code range	Error type
30x, 31x, 32x	SDP keywords
33x	Network services
34x, 35x, 36x	Reserved for future use
37x	QoS parameters
38x	Reserved
39x	Miscellaneous

Table 6.17
SIP Warning Code List

Warning Code	Description
300	Incompatible network protocol
301	Incompatible network address formats
302	Incompatible transport protocol
303	Incompatible bandwidth units
304	Media type not available
305	Incompatible media format
306	Attribute not understood
307	Session description parameter not understood
330	Multicast not available
331	Unicast not available
370	Insufficient bandwidth
399	Miscellaneous Warning

6.3.6 RSeq

The RSeq header [6] is used in a provisional (1xx class) responses to INVITEs to request reliable transport. The header may only be used if the INVITE request contained the Supported: rel100 header. If present in a provisional response, the UAC should acknowledge receipt of the response with a PRACK method, as described in Section 4.1.8. The RSeq header contains a reliable sequence number that is an integer randomly initialized by the UAS. Each subsequent provisional response sent reliably for this call leg will have a monotonically increasing RSeq number. The UAS matches the reliable sequence number and CSeq from the RAck in a PRACK request to a sent response to confirm receipt and stop all retransmissions of the response. An example is:

```
RSeq: 2345263
```

6.4 Entity Headers

Entity headers are used to provide additional information about the message body or the resource requested. This term comes from HTTP [2] where it has a more specific meaning. In SIP, "entity" and "message body" are used interchangably.

Table 6.18
Examples of WWW-Authenticate Header

Header	Meaning
WWW-Authenticate: Digest realm="SIP Telephone Company", domain = "sip-company.com", nonce="9c8e88df84f1cec4341ae6e5a359", opaque="", stale="FALSE", algorithm="MD5"	HTTP digest challenge
WWW-Authenticate: pgp version=5.0; realm="Gateway Password Required"; nonce="76e63aff71"; signature="D2+kk2adfkjfhudgHgFWQncvej"	PGP challenge

6.4.1 Allow

The Allow header is used to indicate the methods supported by the user agent or proxy server sending the response. The header must be present in a 405 Method Not Allowed response and should be included in a positive response to an OPTIONS request. An example is:

```
Allow: INVITE, ACK, BYE, INFO, OPTIONS, CANCEL
```

6.4.2 Content-Encoding

The Content-Encoding header is used to indicate that the listed encoding scheme has been applied to the message body. This allows the UAS to determine the decoding scheme necessary to interpret the message body. Multiple listings in this header indicate that multiple encodings have been used in the sequence in which they are listed. Only encoding schemes listed in an Allow-Encoding header may be used. The compact form is e. Examples include:

```
Content-Encoding: text/plain
e: gzip
```

6.4.3 Content-Disposition

The Content-Disposition header is used to describe the function of a message body. Defined values include session, icon, alert, and render. The value session indicates that the message body contains information to describe a media session. The value render indicates that the message body should be displayed or otherwise rendered for the user. If a message body is present in a request or a 2xx response without a Content-Disposition, the function is assumed to be session. For all other response classes with message bodies, the default function is render. An example is:

```
Content-Function: session
```

6.4.4 Content-Length

The Content-Length is used to indicate the number of octets in the message body. A Content-Length: 0 indicates no message body. As described in Section 2.4.2, this header is used to separate multiple messages sent within a TCP stream. If not present in a request, a 411 Length Required response can be sent. If not present in a UDP message, the message body is assumed to continue to the end of the datagram. If not present in a TCP message, the message body is assumed to continue until the connection is closed. The

Content-Length octet count does not include the CRLF that separates the message headers from the message body. It does, however, include the CRLF at the end of each line of the message body. An example octet calculation is in Chapter 2, footnote 3 (page 18). The Content-Length header is not a required header to allow dynamically generated message bodies, where the Content-Length may not be known a priori. The compact form is l. Examples include:

```
Content-Length: 0
l: 287
```

6.4.5   Content-Type

The Content-Type header is used to specify the Internet media type [9] in the message body. Media types have the familiar form type/sub-type. If this header is not present, application/sdp is assumed. If an Accept header was present in the request, the response Content-Type must contain a listed type, or a 415 Unsupported Media Type response must be returned. The compact form is c. Examples include:

```
Content-Type: application/sdp
c: text/html
```

6.4.6   Expires

The Expires header is used to indicate the time interval in which the request or message contents is valid. When present in an INVITE request, the header sets a time limit on the completion of the INVITE request. That is, the UAC must receive a final response (non-1xx) within the time period or the INVITE request is automatically canceled with a 408 Request Time-out response. Once the session is established, the value from the Expires header in the original INVITE has no effect—the Session-Expires header (Section 6.2.19) must be used for this purpose. When present in a REGISTER request, the header sets the time limit on the URLs in Contact headers that do not contain an expires parameter. Table 4.3 shows examples of the Expires header in registration requests. The header is not defined for any other method types. The header field may contain a SIP date or a number of seconds. Examples include:

```
Expires: 60
Expires: Fri, 15 Apr 2000 00:00:00 GMT
```

6.4.7 MIME-Version

The `MIME-Version` header is used to indicate the version of Multipurpose Internet Mail Extensions Protocol used to construct the message body. SIP, like HTTP, is not considered MIME-compliant because parsing and semantics are defined by the SIP standard, not the MIME Specification [10]. Version 1.0 is the default value. An example is:

```
MIME-Version: 1.0
```

References

[1] Handley, M., et al., "SIP: Session Initiation Protocol," RFC 2543, 1999, Section 6.

[2] Fielding, R., et al.,"Hypertext Transfer Protocol — HTTP/1.1," RFC 2068, June 1999.

[3] Braden, R., "Requirements for Internet Hosts: Application and Support," RFC 1123, 1989.

[4] Rosenberg, J., and H. Schulzrinne, "The SIP Supported Header," IETF Internet-Draft, Work in Progress.

[5] Schulzrinne, H., and J. Rosenberg, "SIP Caller Preferences and Callee Capabilities," IETF Internet-Draft, Work in Progress.

[6] Rosenberg, J., and H. Schulzrinne, "Reliability of Provisional Responses," IETF Internet-Draft, Work in Progress.

[7] Donovan, S., and J. Rosenberg, "The SIP Session Timer," IETF Internet-Draft, Work in Progress.

[8] Elkins, M.,"MIME Security with Pretty Good Privacy (PGP)," RFC 2015, 1996.

[9] Postel, J., "Media Type Registration Procedure," RFC 1590, 1994.

[10] Freed, M., and N. Borenstein, "Multipurpose Internet Mail Extensions (MIME). Part One: Format of Internet Message Bodies," RFC 2045, 1996.

7

Related Protocols

The Session Initiation Protocol (SIP) is one part of the protocol suite that makes up the Internet Multimedia Conferencing architecture as shown in Figure 1.1. In this chapter, other related Internet protocols mentioned or referenced in other sections are introduced, along with details on the use of the protocol with SIP. This is by no means a complete discussion of multimedia communication protocols over the Internet. First, SDP, the media description language, will be discussed. Then the RTP and RTCP media transport protocols will be discussed. The application of RTP/AVP profiles that link SDP and RTP will then be then covered. The chapter concludes with a brief discussion of signaling protocols in the PSTN. The H.323 protocol will be discussed and compared to SIP in the next chapter.

7.1 SDP—Session Description Protocol

The Session Description Protocol, defined by RFC 2327 [1], was developed by the IETF MMUSIC working group. It is more of a description syntax than a protocol in that it does not provide a full-range media negotiation capability. The original purpose of SDP was to describe multicast sessions set up over the Internet's multicast backbone, the MBONE. The first application of SDP was by the experimental Session Announcement Protocol (SAP) [2] used to post and retrieve announcements of MBONE sessions.

121

SAP messages carry a SDP message body, and was the template for SIP's use of SDP. Even though it was designed for multicast, SDP has been applied to the more general problem of describing general multimedia sessions established using SIP.

As seen in the examples of Chapter 3, SDP contains the following information about the media session:

- IP Address (IPv4 address or host name);
- Port number (used by UDP or TCP for transport);
- Media type (audio, video, interactive whiteboard, etc.);
- Media encoding scheme (PCM A-Law, MPEG II video, etc.).

In addition, SDP contains information about the following:

- Subject of the session;
- Start and stop times;
- Contact information about the session.

Like SIP, SDP uses text coding. An SDP message is composed of a series of lines, called fields, whose names are abbreviated by a single lower-case letter, and are in a required order to simplify parsing. The set of mandatory SDP fields is shown in Table 2.1. The complete set is shown in Table 7.1.

SDP was not designed to be easily extensible, and parsing rules are strict. The only way to extend or add new capabilities to SDP is to define a new attribute type. However, unknown attribute types can be silently ignored. A SDP parser must not ignore an unknown field, a missing mandatory field, or an out-of-sequence line. An example SDP message containing many of the optional fields is shown below:

```
v=0
o=johnston 2890844526 2890844526 IN IP4 43.32.1.5
s=SIP Tutorial
i=This broadcast will cover this new IETF protocol
u=http://www.digitalari.com/sip
e=Alan Johnston alan@wcom.com
p=+1-314-555-3333 (Daytime Only)
c=IN IP4 225.45.3.56/236
b=CT:144
t=2877631875 2879633673
```

Table 7.1

SDP Field List in Their Required Order

Field	Name	Mandatory/ Optional
v=	Protocol version number	m
o=	Owner/creator and session identifier	m
s=	Session name	m
i=	Session information	o
u=	Uniform Resource Identifer	o
e=	Email address	o
p=	Phone number	o
c=	Connection information	m
b=	Bandwidth information	o
t=	Time session starts and stops	m
r=	Repeat times	o
z=	Time zone corrections	o
k=	Encryption key	o
a=	Attribute lines	o
m=	Media information	m
a=	Media attributes	o

```
m=audio 49172 RTP/AVP 0
a=rtpmap:0 PCMU/8000
m=video 23422 RTP/AVP 31
a=rtpmap:31 H261/90000
```

The general form of a SDP message is:

```
x=parameter1 parameter2 ... parameterN
```

The line begins with a single lower-case letter x. There are never any spaces between the letter and the =, and there is exactly one space between each parameter. Each field has a defined number of parameters. Each line ends with a CRLF. The individual fields will now be discussed in detail.

7.1.1 Protocol Version

The v= field contains the SDP version number. Because the current version of SDP is 0, a valid SDP message will always begin with v=0.

7.1.2 Origin

The o= field contains information about the originator of the session and session identifiers. This field is used to uniquely identify the session. The field contains:

```
o=username session-id version network-type address-type
address
```

The username parameter contains the originator's login or host or - if none. The session-id parameter is a Network Time Protocol (NTP) [3] timestamp or a random number used to ensure uniqueness. The version is a numeric field that is increased for each change to the session, also recommended to be a NTP timestamp. The network-type is always IN for Internet. The address-type parameter is either IP4 or IP6 for IPv4 or IPv6 address either in dotted decimal form or a fully qualified host name.

7.1.3 Session Name and Information

The s= field contains a name for the session. It can contain any non-zero number of characters. The optional i= field contains information about the session. It can contain any number of characters.

7.1.4 URI

The optional u= field contains a uniform resource indicator (URI) with more information about the session.

7.1.5 E-mail Address and Phone Number

The optional e= field contains an e-mail address of the host of the session. If a display name is used, the e-mail address is enclosed in <>. The optional p= field contains a phone number. The phone number should be given in globalized format, beginning with a +, then the country code, a space or -, then the local number. Either spaces or - are permitted as spacers in SDP. A comment may be present in ().

7.1.6 Connection Data

The c= field contains information about the media connection. The field contains:

```
c=network-type address-type connection-address
```

The network-type parameter is defined as IN for the Internet. The address type is defined as IP4 for IPv4 addresses. The connection-address is the IP address that will be sending the media packets, which could be either multicast or unicast. If multicast, the connection-address field contains:

```
connection-address=base-multicast-address/ttl/number-of-
addresses
```

where ttl is the time-to-live value, and number-of-addresses indicates how many contiguous multicast addresses are included starting with the base-multicast-address.

7.1.7 Bandwidth

The optional b= field contains information about the bandwidth required. It is of the form:

```
b=modifier:bandwidth-value
```

The modifier is either CT for conference total or AS for application specific. CT is used for multicast session to specify the total bandwidth that can be used by all participants in the session. AS is used to specify the bandwidth of a single site. The bandwidth-value parameter is the specified number of kilobytes per second.

7.1.8 Time, Repeat Times, and Time Zones

The t= field contains the start time and stop time of the session.

```
t=start-time stop-time
```

The times are specified using NTP timestamps. For a scheduled session, a stop-time of zero indicates that the session goes on indefinitely. A start-time and stop-time of zero for a scheduled session indicates that it is permanent. The optional r= field contains information about the repeat times that can be specified in either in NTP or in days (d), hours (h), or minutes (m). The optional z= field contains information about the time zone offsets. This field is used if a reoccurring session spans a change from daylight-savings to standard time, or vice versa.

7.1.9 Encryption Keys

The optional k= field contains the encryption key to be used for the media session. The field contains:

    k=method:encryption-key

The method parameter can be clear, base64, uri, or prompt. If the method is prompt, the key will not be carried in SDP; instead, the user will be prompted as they join the encrypted session. Otherwise, the key is sent in the encryption-key parameter.

7.1.10 Media Announcements

The optional m= field contains information about the type of media session. The field contains:

    m=media port transport format-list

The media parameter is either audio, video, application, data, or control. The port parameter contains the port number. The transport parameter contains the transport protocol, which is either RTP/AVP or udp. (RTP/AVP stands for Real-time Transport Protocol [4] / audio video profiles [5], which is described in Section 7.3.) The format-list contains more information about the media. Usually, it contains media payload types defined in RTP audio video profiles. More than one media payload type can be listed, allowing multiple alternative codecs for the media session. For example, the following media field lists three codecs:

    m=audio 49430 RTP/AVP 0 6 8

One of these three codecs can be used for the audio media session. If the intention is to establish three audio channels, three separate media fields would be used. For non-RTP media, Internet media types should be listed in the `format-list`. For example,

```
m=application 52341 udp wb
```

could be used to specify the `application/wb` media type.

7.1.11 Attributes

The optional `a=` field contains attributes of the preceding media session. This field can be used to extend SDP to provide more information about the media. If not fully understood by a SDP user, the attribute field can be ignored. There can be one or more attribute fields for each media payload type listed in the media field. For the RTP/AVP example in Section 7.1.10, the following three attribute fields could follow the media field:

```
a=rtpmap:0 PCMU/8000
a=rtpmap:6 DVI4/16000
a=rtpmap:8 PCMA/8000
```

Other attributes are shown in Table 7.2. Full details of the use of these attributes are in the standard document [1].

7.1.12 Use of SDP in SIP

The default message body type in SIP is `application/sdp`. The calling party lists the media capabilities that they are willing to receive in SDP in either an `INVITE` or in an `ACK`. The called party lists their media capabilities in the `200 OK` response to the `INVITE`.

Because SDP was developed with scheduled multicast sessions in mind, many of the fields have little or no meaning in the context of dynamic sessions established using SIP. In order to maintain compatibility with the SDP protocol, however, all required fields are included. A typical SIP use of SDP includes the version, origin, subject, time, connection, and one or more media and attribute fields as shown in Table 2.1. The origin, subject, and time fields are not used by SIP but are included for compatibility. In the SDP standard, the subject field is a required field and must contain at least one character, suggested to be `s=-` if there is no subject. The SIP standard,

Table 7.2
SDP Attribute values

Attribute	Name
a=rtpmap:	RTP/AVP list
a=cat:	Category of the session
a=keywds:	Keywords of session
a=tool:	Name of tool used to create SDP
a=ptime:	Length of time in milliseconds for each packet
a=recvonly	Receive only mode
a=sendrecv	Send and receive mode
a=sendonly	Send only mode
a=orient:	Orientation for whiteboard sessions
a=type:	Type of conference
a=charset:	Character set used for subject and information fields
a=sdplang:	Language for the session description
a=lang:	Default language for the session
a=framerate:	Maximum video frame rate in frames per second
a=quality:	Suggests quality of encoding
a=fmtp:	Format transport

however, allows the subject field to be omitted for two-party sessions. The time field is usually set to t=0 0.

SIP uses the connection, media, and attribute fields to set up sessions between user agents. Because the type of media session and codec to be used are part of the connection negotiation, SIP can use SDP to specify multiple alternative media types and to selectively accept or decline those media types. When multiple media codecs are listed, the caller and called party's media fields must be aligned—that is, there must be the same number, and they must be listed in the same order. The SIP standard recommends that an attribute containing a=rtpmap: be used for each media field [6]. A media stream is declined by setting the port number to zero for the corresponding media field in the SDP response. In the following example, the caller Tesla wants to set up an audio and video call with two possible audio codecs and a video codec in the SDP carried in the initial INVITE:

```
v=0
o=Tesla 2890844526 2890844526 IN IP4 lab.high-voltage.org
s=-
c=IN IP4 100.101.102.103
t=0 0
m=audio 49170 RTP/AVP 0 8
a=rtpmap:0 PCMU/8000
a=rtpmap:8 PCMA/8000
m=video 49172 RTP/AVP 32
a=rtpmap:32 MPV/90000
```

The codecs are referenced by the RTP/AVP profile numbers 0, 8, and 32. The called party Marconi answers the call, chooses the second codec for the first media field and declines the second media field, only wanting a PCM A-Law audio session.

```
v=0
o=Marconi 2890844526 2890844526 IN IP4 tower.radio.org
s=-
c=IN IP4 200.201.202.203
t=0 0
m=audio 60000 RTP/AVP 8
a=rtpmap:8 PCMA/8000
m=video 0 RTP/AVP 32
```

If this audio-only call is not acceptable, then Tesla would send an ACK then a BYE to cancel the call. Otherwise, the audio session would be established and RTP packets exchanged. As this example illustrates, unless the number and order of media fields is maintained, the calling party would not know for certain which media sessions were being accepted and declined by the called party.

One party in a call can temporarily place the other on hold (i.e., suspending the media packet sending). This is done by sending an INVITE with identical SDP to that of the original INVITE but with the IP address set to 0.0.0.0 in the c= field. The call is made active again by sending another INVITE with the IP address set back to that of the user agent.

7.2 RTP—Real-time Transport Protocol

Real-time Transport Protocol [4] was developed to enable the transport of real-time packets containing voice, video, or other information over IP. RTP is defined by IETF Proposed Standard RFC 1889. RTP does not provide any quality of service over the IP network—RTP packets are handled the same as all other packets in an IP network. However, RTP allows for the detection of some of the impairments introduced by an IP network, such as:

- packet loss;
- variable transport delay;
- out of sequence packet arrival;
- asymmetric routing.

As shown in the protocol stack of Figure 1.1, RTP is an application layer protocol that uses UDP for transport over IP. RTP is not text encoded, but uses a bit-oriented header similar to UDP and IP. RTP version 0 is only used by the *vat* audio tool for MBONE broadcasts. Version 1 was a pre-RFC implementation and is not in use. The current RTP version 2 packet header is shown in Figure 7.1. RTP was designed to be very general; most of the headers are only loosely defined in the standard; the details are left to profile documents. The 12 octets are defined as:

- Version (V). This 2-bit field is set to 2, the current version of RTP.
- Padding (P). If this bit is set, there are padding octets added to the end of the packet to make the packet a fixed length. This is most commonly used if the media stream is encrypted.
- Extension (X). If this bit is set, there is one additional extension following the header (giving a total header length of 14 octets). Extensions are defined by certain payload types.
- CSRC count (CC). This 4-bit field contains the number of content source identifiers (CSRC) are present following the header. This

V	P	X	CC	M	PT	Sequence Number	Timestamp	SSRCI

Figure 7.1 RTP packet header.

field is only used by mixers that take multiple RTP streams and output a single RTP stream.

- Marker (M). This single bit is used to indicate the start of a new frame in video, or the start of a talk-spurt in silence-suppressed speech.

- Payload Type (PT). This 7 bit field defines the codec in use. The value of this field matches the profile number listed in the SDP.

- Sequence Number. This 16-bit field is incremented for each RTP packet sent and is used to detect missing/out of sequence packets.

- Timestamp. This 32-bit field indicates in relative terms the time when the payload was sampled. This field allows the receiver to remove jitter and to play back the packets at the right interval assuming sufficient buffering.

- Synchronization Source Identifier (SSRCI). This 32-bit field identifies the sender of the RTP packet. At the start of a session, each participant chooses a SSRC number randomly. Should two participants choose the same number, they each choose again until each party is unique.

- CSRC Contributing Source Identifier. There can be none or up to 15 instances of this 32-bit field in the header. The number is set by the CSRC Count (CC) header field. This field is only present if the RTP packet is being sent by a mixer, which has received RTP packets from a number of sources and sends out combined packets. A non-multicast conference bridge would utilize this header.

RTP allows detection of a lost packet by a gap in the Sequence Number. Packets received out of sequence can be detected by out-of-sequence Sequence Numbers. Note that RTP allows detection of these transport-related problems but leaves it up to the codec to deal with the problem. For example, a video codec may compensate for the loss of a packet by repeating the last video frame, while an audio codec may play background noise for the interval. Variable delay or jitter can be detected by the Timestamp field. A continuous bit rate codec such as PCM will have a linearly increasing Timestamp. A variable bit rate codec, however, which sends packets at irregular intervals, will have an irregularly increasing Timestamp, which can be used to play back the packets at the correct interval.

The RTP Control Protocol (RTCP) is a related protocol also defined in RFC 1889 that allows participants in an RTP session to send each other quality reports and statistics, and exchange some basic identity information. The four types of RTCP packets are shown in Table 7.3. RTCP has been designed to scale to very large conferences. Because RTCP traffic is all overhead, the bandwidth allocated to these messages remains fixed regardless of the number of participants. That is, the more participants on a conference, the less frequently RTCP packets are sent. For example, in a basic two-participant audio RTP session, the RTP/AVP profile states that RTCP packets are to be sent about every 5 seconds; for four participants, RTCP packets can be sent every 10 seconds. Sender reports (SR) or receiver reports (RR) packets are sent the most frequently, with the other packet types being sent less frequently. The use of reports allows feedback on the quality of the connection including information such as:

- number of packets sent and received;
- number of packets lost;
- packet jitter.

In a multimedia session established with SIP, the information needed to select codecs and send the RTP packets to the right location is carried in the SDP message body. Under some scenarios, it can be desirable to change codecs during an RTP session. An example of this relates to the transport of dual tone multiple frequency (DTMF) digits. A low bit rate codec that is optimized for transmitting vocal sounds will not transport the superimposed sine waves of a DTMF signal without introducing significant noise, which

Table 7.3
RTCP Packet Types

Packet type	Name	Description
SR	Sender report	Sent by a participant that both sends and receives RTP packets
RR	Receiver report	Sent by a participant that only receives RTP packets
SDES	Source description	Contain information about the participant in the session including e-mail address, phone number, and host
BYE	Bye	Sent to terminate the RTP session
APP	Application specific	Defined by a particular profile

may cause the DTMF digit receiver to fail to detect the digit. As a result, it is useful to switch to another codec when the sender detects a DTMF tone. Because a RTP packet contains the payload type, it is possible to change codecs "on the fly" without any signaling information being exchanged between the user agents. On the other hand, switching codecs in general should probably not be done without a SIP signaling exchange (re-INVITE) beacuse the call could fail if one side switches to a codec that the other does not support. The SIP re-INVITE message exchange allows this change in media session parameters to fail without causing the established session to fail.

The use of random numbers for CSRC provides a minimal amount of security against "media spamming" where a literally *uninvited* third party tries to break into a media session by sending RTP packets during an established call. Unless the third party can guess the CSRC of the intended sender, the receiver will detect a change in CSRC number and either ignore the packets or inform the user that something is going on. This behavior for RTP clients, however, is not universally accepted, because in some scenarios (wireless hand-off, announcement server, call center, etc.) it might be desirable to send media from multiple sources during the progress of a call.

RTP supports encryption of the media. In addition, RTP can use IPSec [7] for authentication and encryption.

7.3 RTP Audio Video Profiles

The use of profiles enables RTP to be an extremely general media transport protocol. The current audio video profiles defined by RFC 1890 are listed in Table 7.4. The profile document makes the following specifications for RTP:

- UDP is used for underlying transport;
- RTP port numbers are always even, the corresponding RTCP port number is the next highest port, always an odd number;
- No header extensions are used.

For each of the profiles listed in Table 7.4, the profile document lists details of the codec, or a reference for the details is provided. Payloads in the range 96–127 can be defined dynamically during a session. The minimum payload support is defined as 0 (PCMU) and 5 (DVI4). The document recommends dynamically assigned port numbers, although 5004 and 5005 have been registered for use of the profile and can be used instead. The standard also describes the process of registering new payload types with IANA. There are other references for a tutorial description of many of these audio codecs [8] and video codecs [9].

The information in the first three columns of Table 7.4 is also contained in the SDP `a=rtpmap:` field, which is why the attribute is optional.

Table 7.4
RTP/AVP Audio and Video Payload Types

Payload	Codec	Clock	Description
0	PCMU	8000	ITU G.711 PCM µ-Law Audio 64kbps
1	1016	8000	CELP Audio 4.8kbps
2	G721	8000	ITU G721 ADPCM Audio 32kbps
3	GSM	8000	European GSM Audio 13kbps
5	DVI4	8000	DVI ADPCM Audio 32kbps
6	DVI4	16000	DVI ADPCM 64kbps
7	LPC	8000	Experimental LPC Audio
8	PCMA	8000	ITU G.711 PCM A-Law Audio 64kbps
9	G722	8000	ITU G.722 Audio
10	L16	44100	Linear 16 bit Audio 705.6kbps
11	L16	44100	Linear 16 bit Stereo Audio 1411.2kbps
14	MPA	90000	MPEG-I or MPEG-II Audio Only
15	G728	8000	ITU G.728 Audio 16kb/s
25	CELB	90000	CelB Video
26	JBEG	90000	JBEG Video
28	NV	90000	nv Video
31	H261	90000	ITU H.261 Video
32	MPV	90000	MPEG-I and MPEG-II Video
33	MP2T	90000	MPEG-II transport stream Video

7.4 PSTN Protocols

Three types of PSTN signaling protocols are mentioned in this text: Circuit Associated Signaling (CAS), ISDN (Integrated Services Digital Network), and ISUP (ISDN User Part). They will be briefly introduced and explained. How these protocols work in the PSTN today are covered in other references [8].

7.4.1 Circuit Associated Signaling

This type of signaling is the oldest currently used in the PSTN today. The signaling information uses the same audio circuit as the voice path, with digits and characters represented by audio tones. These are the tones that used to be barely discernible on long-distance calls before ring tone is heard. The tones are called multi-frequency (MF) tones. They are similar to the tones used to signal between a telephone and a central office switch, which are DTMF tones. Long post dial delay is introduced because of the time taken to out-pulse long strings of digits. Also, CAS is susceptible to fraud, as fraudulent tones can be generated by the caller to make free telephone calls. This type of signaling is common in trunk circuits between a central office and a corporation's private branch exchange (PBX) switch.

7.4.2 ISUP Signaling

ISDN User Part is the protocol used between telephone switches in the PSTN for call signaling. It is used over a dedicated packet-switched network that uses Signaling System #7 (SS7) for transport. This signaling method was developed to overcome some of the delay and security problems with CAS. There are examples of ISUP signaling in the call flow examples of Chapter 9. The adoption of this out-of-band signaling protocol was the first step taken by telecommunications carriers away from circuit-switched networks and towards packet-switched networks. The final step will be moving the bearer channels onto a packet-switched network.

7.4.3 ISDN Signaling

Integrated Services Digital Network signaling was developed for all-digital telephone connections to the PSTN. The most common types of interfaces are the basic rate interface (BRI) and the primary rate interface (PRI). A BRI can contain two 64-kbps B-channels for either voice or data and a 16 kbps D-channel for signaling. BRI can be used over conventional telephone lines but requires an ISDN telephone or terminal adapter. PRI uses a 1.544-Mbps link called a T-1 or a DS-1, which is divided up into 23 B-channels and one D-channel, with each channel being 64 kbps. The H.323 protocol, described in Chapter 9, reuses a subset of the ISDN Q.931 signaling protocol used over the D-channel.

References

[1]   Handley, M., and V. Jacobson, "SDP: Session Description Protocol," RFC 2327, 1998.

[2]   Handley, M., C. Perkins, and E. Whelan, "Session Announcement Protocol," Internet-Draft, Work in Progress.

[3]   Mills, D., "Network Time Protocol (Version 3): Specification, Implementation, and Analysis," RFC 1305, 1992

[4]   Schulzrinne, H., et al., "RTP: A Transport Protocol for Real-time Applications," RFC 1889, 1996.

[5]   Schulzrinne, H., "RTP Profile for Audio and Video Conferences with Minimal Control," RFC 1890, 1996.

[6]   Handley, M., et al., "SIP: Session Initiation Protocol," RFC 2543, 1999, Appendix B.

[7]   Kent, S., and R. Atkinson, "Security Architecture for the Internet Protocol," RFC 2401, 1998.

[8]   Anttalainen, T., *Introduction to Telecommunications Network Engineering*, Artech House: Norwood, MA, 1999.

[9]   Schaphorst, R., *Videoconferencing and Videotelephony: Technology and Standards*, 2nd Ed., Artech House: Norwood, MA, 1999.

8

Comparison to H.323

This chapter compares SIP to another IP telephony signaling protocol: the International Telecommunications Union (ITU) recommendation H.323, entitled "Packet-based Multimedia Communication." H.323 is introduced and explained using a simple call flow example. H.323 and SIP are then compared.

8.1 Introduction to H.323

H.323 [1] is an umbrella recommendation that covers all aspects of multimedia communication over an IP network. It is part of the H.32x series[1] of protocols that describes multimedia communication over ISDN, broadband (ATM)[2], telephone (PSTN), and packet (IP) networks, as shown in Table 8.1. Originally developed for video conferencing over a single LAN

---

1. In this chapter, the use of an x instead of a digit does not imply that all digits (0–9) in the range are included. In this case H.32x does not include H.325 to H.329, which have yet to be defined.

2. In this context, broadband means transported over an Asynchronous Transfer Mode (ATM) network. In anticipation of the universal deployment of ATM networks by carriers, the ITU developed a suite of protocols to support conventional telephony over ATM networks. For example, Q.2931 is the extension of Q.931 ISDN over ATM. Today, the term is used to mean high bandwidth connections—faster than modem speeds.

Table 8.1
ITU H.32x Family of Standards

Protocol	Title
H.320	Communication over ISDN networks
H.321	Communication over broadband ISDN (ATM) networks
H.322	Communication over LANs with guaranteed QoS
H.323	Communication over LANs with non-guaranteed QoS (IP)
H.324	Communication over PSTN (V.34 modems)

segment, the protocol has been extended to cover the general problem of telephony over the Internet. The first version was approved by the ITU in 1996, and it was adopted by early IP telephony networks because there were no other standards. Version 2 was adopted in 1998 to fix some of the problems and limitations in version 1. Version 3 was adopted in 1999 and includes modifications and extensions to enable communications over a larger network. H.323 has been designed to be backward compatible, so a version 1 compliant terminal can communicate with a version 3 gatekeeper.

H.323 references a number of other ITU and IETF protocols to completely specify the environment. Each element of the network is defined and standardized. Figure 8.1 shows the main elements: terminals, gatekeepers, gateways, and multipoint control units (MCUs). An H.323 terminal is an end device in the network. It originates and terminates media streams that could be audio, video, or data, or a combination of all three. At a minimum, all H.323 terminals must support basic G.711 PCM audio transmission. Support of video and data are optional. An H.323 gatekeeper is a server that controls a zone, which is an administrative domain in H.323. If a gatekeeper is present, all terminals within that zone must register with and defer to the gatekeeper on authorization decisions to place or accept a call. A gatekeeper also provides services to terminals in a zone, such as gateway location, address translation, bandwidth management, feature implementation, and registration. A gatekeeper is not a required element in an H.323 network, but a terminal's capabilities without one are severely limited. A gateway is another optional element in an H.323 network. It interfaces the H.323 network with another protocol network, such as the PSTN. An MCU provides

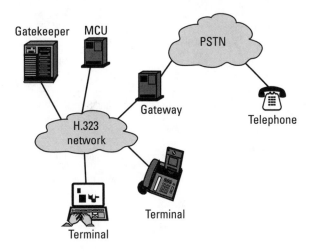

Figure 8.1 Elements of an H.323 network.

conferencing services for terminals, which are restricted to small conferences of three participants without a MCU.

Some of the protocols referenced by H.323 are shown in Table 8.2. H.225 is used for registration, admission, and status (RAS), which is used for terminal to gatekeeper communication. A subset of Q.931 is used for call setup signaling between terminals. H.245 is used for control signaling or media negotiation and capability exchange between terminals. T.120 is used for multipoint graphic communications. H.323 audio codecs are specified in the ITU G.7xx series. Video codecs are specified in the H.26x series. H.323 also references two IETF protocols, RTP and RTCP, for the media transport. The H.235 recommendation covers privacy and encryption, while H.450 covers supplementary services such as those commonly found in the PSTN (e.g., call forwarding, call hold, call park, etc.).

The protocols referenced by H.323 are not suggestions; they are requirements. For example, only ITU approved and standardized codecs (G.7xx) can be used by H.323 terminals.

8.2 Example of H.323

Figure 8.2 shows a simple call flow involving two terminals and a gatekeeper. The call begins with an exchange of H.225 RAS messages between the

Table 8.2
Protocols Referenced by H.323

Protocol	Description
H.225	Registration, Admission, and Status (RAS)
Q.931	Call Signaling
H.245	Control Signaling (Media Negotiation)
T.120	Multipoint Graphic Communication
G.7xx	Audio Codecs
H.26x	Video Codecs
RTP	Real-time Transport Protocol (RTP)
RTCP	RTP Control Protocol (RTCP)
H.235	Privacy and Encryption
H.450	Supplementary Services

calling terminal and the gatekeeper. It is assumed that both terminals have already registered with the gatekeeper using the Registration Request (RRQ) message. The calling terminal opens a TCP connection to the gatekeeper and sends an Admission Request (ARQ) message to the gatekeeper containing the address of the called terminal and the type of session desired. The address could be specified as an H.323 alias, E.164 telephone number, e-mail address, or a URL. The gatekeeper knows about all calls in the zone it controls; it decides if the user is authorized to make a call and if there is enough bandwidth or other resources available. In this example, there is enough bandwidth, so the gatekeeper allows the call to continue by sending an Admission Confirmation (ACF) message. The ACF indicates to the calling terminal that end-point message routing, or the direct exchange of H.225 call signaling messages with the called terminal, is to be used. Alternatively, the gatekeeper can require gatekeeper routed signaling, where the gatekeeper acts like a proxy and forwards all messages between the terminals. The gatekeeper has also translated the address in the ARQ into an IP address that was returned in the ACF. The RAS TCP connection to the gatekeeper can now be closed.

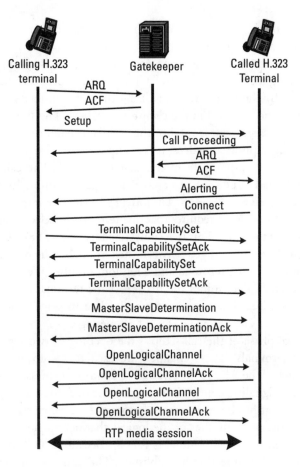

Figure 8.2 H.323 call flow example.

The calling terminal is now able to open a TCP connection to the called terminal using the well-known H.225 port number 1720 and send a Q.931 Setup message to the called terminal. The called terminal responds with a Call Proceeding response to the calling terminal. The called terminal must also get permission from the gatekeeper before it accepts the call, so an ARQ is sent to the gatekeeper. When it receives the ACF from the gatekeeper, the called terminal begins alerting the user, and sends an Alerting message to the calling terminal. When the user at the calling terminal answers, a Connect message is sent. There is no acknowledgment of messages because all these messages are sent using TCP, which provides reliable transport. These call signaling messages used in H.323 are a subset of the Q.931 recommendation that covers ISDN D-channel signaling.

The next stage is the connection negotiation, which is handled by H.245 control signaling messages. A second TCP connection between the two terminals is opened by the calling terminal using the port number selected by the called terminal and returned in the Connect message. The TerminalCapabilitySet message sent contains the media capabilities of the calling terminal, listing supported codecs. It is acknowledged with a TerminalCapabilitySetAck response from the called terminal. The called terminal then sends a TerminalCapabilitySet message containing its media capabilities, which receives a TerminalCapabilitySetAck response.

The H.323 protocol requires that one terminal be selected as the master with the other as the slave. This is accomplished using MasterSlaveDetermination messages exchanged between the terminals. The messages contain the terminal type of the terminal and a random number. Terminal types are hierarchical, which determines the master. If the terminal type is the same, the random number determines the master. The message is acknowledged with a MasterSlaveDeterminationAck message.

The final phase of the call setup is the opening of two logical channels between the terminals. These channels are used to set up and control the media channels. The H.245 OpenLogicalChannel message sent in the H.245 control signaling connection contains the desired media type, including the codec that has been determined from the exchange of capabilities. It also contains the port number of the logical channel opened to control the media. The media control channel is RTCP. Because RTCP uses UDP for transport, no TCP connection is established. The terminal, however, will listen on that port for UDP datagrams. The message is acknowledged with an OpenLogicalChannelAck message.

Now, the terminals begin sending RTP media packets and also RTCP control packets using the IP addresses and port numbers exchanged in the OpenLogicalChannel messages.

Figure 8.3 shows a call tear-down sequence, which either terminal may initiate. In this example, the called terminal sends an EndSessionCommand message in the H.245 control signaling channel. The other terminal responds with an EndSessionCommand message in the H.245 control signaling channel, which can now be closed. The called terminal then opens an RAS TCP connection to the gatekeeper (unless one is already open) and sends a Disengage Request (DRQ) message and receives a Disengage Confirmation (DCF) message from the gatekeeper. This way, the gatekeeper knows that the resources used in the call have now been freed up. A Call Detail

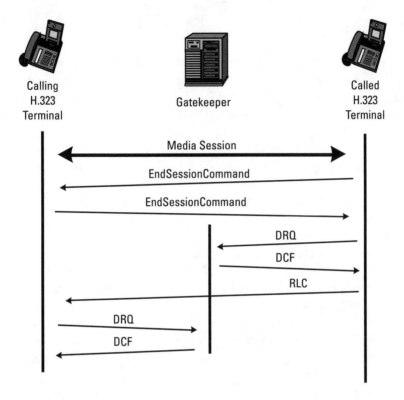

Figure 8.3 H.323 call tear-down sequence.

Record (CDR) or other billing record can be written and stored by the gatekeeper. Next, a Q.931 Release Complete (RLC) message is sent in the call signaling connection, which can then be closed. Finally, the other terminal opens a RAS TCP connection to the gatekeeper (unless one is already open) and sends a DRQ to the gatekeeper and receives a DCF response. The RAS TCP connection can then be closed.

The call flows in Figures 8.2 and 8.3 show direct end-point signaling, where the calling terminal opens TCP connections to the called terminal and exchanges H.225 and H.245 messages. In the ACF response to the calling terminal, the gatekeeper can require gatekeeper routed signaling, where the call signaling and control signaling channels are opened with the gatekeeper, who then opens the channels with the called terminal. In this way, the gatekeeper stays in the signaling path and proxies all messages. This allows the gatekeeper to know the exact call state and be able to invoke features.

8.3 Versions

There are three versions of H.323, which reflect the evolution of this protocol. H.323 is fully backward compatible, so gatekeepers and terminals must support flows and mechanisms defined in all three versions. Version 1 was approved in 1996 and was titled "Visual Telephone Systems over Networks with Non-guaranteed Quality of Service". The example call flow in Figure 8.2 shows the version 1 call setup. Not unexpectedly given the number of messages and TCP connections, this process was very slow, sometimes taking 30 seconds or more to establish a call. While this may have been acceptable for a protocol designed for video conferencing over a single LAN segment, it is not acceptable for an IP telephony network designed to provide a similar level of service to the PSTN.

Version 2 included alternative call setup schemes to speed up the call setup. Two schemes were added to H.323, called FastStart and H.245 tunneling. FastStart is shown in Figure 8.4, in which the *Setup* message contains the TerminalCapabilitySet information. This saves multiple messages and round trips. In H.245 tunneling, a separate H.245 control channel is not opened. Instead, H.245 messages are encapsulated in Q.931 messages in the call signaling channel. This saves overhead in opening and closing a second TCP connection.

Version 3 of H.323 introduced some more functionality to gatekeepers. In the earlier versions, calling was generally assumed to be limited to a single H.323 zone or through a gateway to the PSTN. H.323 did not define a standard for calls that needed to be routed outside of a zone through another gatekeeper. Version 3 introduced gatekeeper-to-gatekeeper signaling using H.225 RAS messages. Figure 8.5 shows a call setup from one zone to another showing version 3 gatekeeper-to-gatekeeper signaling. In this example, gatekeeper 1 forwards the ARQ to gatekeeper 2 who provides the IP address of the called party and allows the call to proceed with an ACF to gatekeeper 1. If gatekeeper 2 cannot locate the called party, or otherwise does not permit the call to proceed, a Admission Denial (ADN) is sent instead. A similar exchange of DRQ messages is required on call tear-down.

8.4 Comparison

This section will compare H.323 with SIP in the following areas: encoding, transport, addressing, complexity, feature implementations, vendor support, conferencing, and extensibility.

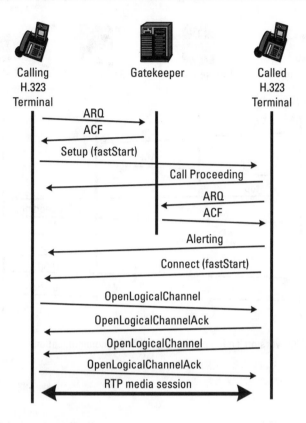

Figure 8.4 H.323 version 2 FastStart call setup.

8.4.1 Encoding

H.323 uses Abstract Syntax Notation 1 (ASN.1) to represent the protocol. Messages are encoded using packed encoding rule (PER). This binary encoding scheme is used to minimize the number of bits required to transport a given field. This is achieved by setting very tight conditions on what a particular field may contain in terms of characters and length, and by compressing the resulting string. For example, an H.323 alias address is limited to 256 characters. In comparison, there is no limit to the number of characters in SIP URLs[3]. An ASN.1 PER encoder and decoder stack is required on every H.323 device, which adds to the complexity. In addition, any test sets used for troubleshooting or monitoring must have specific software to decode a H.323 message—otherwise all that can be displayed is a hex dump that needs to be broken down by hand. Test sets for other ITU protocols such as SS7,

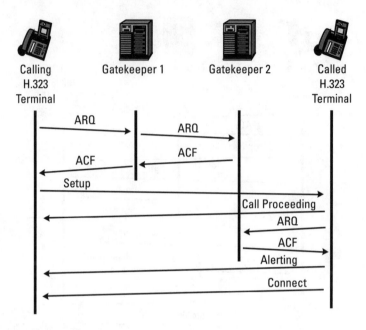

Figure 8.5 H.323 version 3 gatekeeper-to-gatekeeper communication

ISUP, and ISDN are expensive compared to the sets needed for a text-encoded protocol.

SIP uses Augmented Backus Naur Format (ABNF) for representation, and text-based encoding, where the contents of each field are extremely flexible and loosely defined. For example, few headers in SIP have a fixed length. Parser logic is borrowed from HTTP parsers and can be implemented in simple text languages such as Perl. In addition, any packet sniffer can display a SIP message (unless it is encrypted) since the headers and parameters are in plain text.

---

3. This type of limitation on addressing may seem reasonable, but development of advanced services in the Internet has shown otherwise. For example, the use of CGI scripting in URLs in the Internet has proven to be extremely powerful for web sites, which often results in URLs much longer than this H.323 limitation. For example, searching for ietf sip at the "Northern Light" search engine generates a customized search results page with the following URL that has 184 characters: http://www.northernlight.com/nlquery.fcg?ho= groucho&po=5125&qr=ietf+sip&cb=0&ccfor-IP+%28Internet+Protocol%29&cl=1& clsub_1_0=20096& cn_1=IP+%28Internet+Protocol%29&db=97044288&orl=

8.4.2 Transport

H.323 requires TCP for reliable message transport. Since H.323 does not have timers and message retransmission schemes, UDP can not be used. H.323 opens multiple TCP connections to establish a session. A connection must be opened with the gatekeeper for the exchange of H.225 RAS messages. A connection must be opened with the other terminal for the exchange of Q.931 call signaling messages. Finally, one H.245 call control connection must be established for each media session. The H.225 channels can be closed once the call is established (although they must then be opened again to terminate the call), but the H.245 control channels must remain open for the duration of the call. H.245 tunneling in version 2 allows the use of a single TCP connection but at a cost of additional processing.

TCP introduces additional round-trip delays in H.323 message exchanges due to the handshake SYN/ACK exchange during the opening of the TCP connection, as shown in Figure 1.2. In addition, if the setup message is larger than the MTU of the network (maximum packet size), it must be sent in two TCP segments. The slow-start algorithm of TCP requires an ACK to be received for the first segment before the second segment can be sent, causing an additional round-trip delay. There is discussion about modifying H.323 to run over UDP, which may occur in another version of the protocol.

SIP can use either UDP or TCP for transport. If SIP uses TCP, a single TCP connection is established from the UAC to the UAS or proxy. The connection can be closed after the call is established, then reopened again to terminate the call. If the simpler UDP transport is used, no connections need be opened or closed—datagrams are simply sent as needed, and lost messages are handled by SIP. Most SIP implementations have chosen UDP over TCP for efficient and low-latency transport.

8.4.3 Addressing

H.323 addressing uses a number of schemes. Addresses include H.323 aliases, e-mail addresses, phone numbers, URLs, and others. Unless an H.323 terminal knows the IP address of the called terminal, a gatekeeper is required to resolve the address.

The use of URLs by H.323 might seem similar to SIP; however, in H.323, the URL only references a location (the protocol is always assumed to be H.323). SIP can even use H.323 URLs to locate an H.323 terminal, which would then be called using H.323. SIP, through the use of DNS queries, can place calls using SIP URLs without requiring a proxy server to do address resolution.

8.4.4 Complexity

A quick comparison of the complexity of SIP and H.323 can be seen by comparing the call flows of Figure 2.2 for SIP and Figure 8.2 for H.323. The message count of 4 versus 18 sent to set up a call is indicative. Others have compared page counts in the standards documents (178 versus 736) [2], although this is not always a fair comparison criteria!

The choice of signaling protocol also has other less obvious impacts on complexity. The more demands a protocol makes on a system's storage and processing resources, the more expensive the system will be. This complexity is sometimes discussed in terms of scalability, which refers to the ability of a network to grow from a handful of users to hundreds of thousands of users without running into major problems, limitations, or cost barriers, or the ability to build palm-sized basic devices all the way up to huge servers. The amount of call state information, the number of messages processed per call, and the complexity of encoding and decoding messages all play a critical role in the scalability of a protocol.

H.323 was originally developed for use over a single LAN segment. As a result, it has many scalabilty issues. Some of these issues have been addressed in later versions of the protocol, but others remain. The biggest problem relates to the scalability of a gatekeeper, of which there can only be one in a zone. The call state and connection information stored for each call limits the number of simultaneous calls through an H.323 zone. Also, important areas such as message loop detection can only be implemented in a stateful way.

SIP design was based largely on HTTP, which is an example of an extremely successful IETF protocol that has successfully survived years of exponential growth. Many implementers expect that SIP will also undergo a similar growth rate with similar positive results. The use of UDP transport and stateless proxies should allow for more dense servers and gateways than H.323. SIP's use of stateless techniques for message loop detection and other proven scalable protocols such as DNS should also add to its scalability.

8.4.5 Feature Implementations

Both SIP and H.323 can implement features common to the PSTN network such as call forwarding and call transfer. H.323 specifies details of these features in the H.450 supplementary services recommendation. Because the IETF standardizes protocols but not services, the SIP standard does not specify how the features are to be implemented. Some PSTN-like features in SIP require some additional extensions (described in Chapter 10) that are still being developed.

The main difference between SIP and H.323 relates to non-PSTN features. Because SIP uses Internet URLs and places no limitations on the type and number of sessions that it can be used to establish, many new advanced services and features on the Internet will likely make use of SIP rather than H.323.

8.4.6 Vendor Support

The 3-year head start that H.323 has over SIP is reflected in the marketplace today. Most commercial Voice-over-IP implementations in early 2000 use H.323 for signaling. Nearly all vendors of H.323 products, however, are also developing SIP products that will come to market starting in 2000. In addition, there are startup companies that are developing SIP-only products to enable advanced services. The list of vendors participating in the bake-off interoperability tests shows some of the depth of support that SIP is gaining in the marketplace.

8.4.7 Conferencing

H.323 terminals are permitted to perform conferencing with only three other parties. Conferences larger than three parties require the use of a multipoint conference unit (MCU). The MCU receives media streams from each participant and mixes them together on a single stream. The MCU also has built in conference management features, such as floor and microphone control, and conference statistics.

SIP places no restrictions on the number of parties with which a SIP user agent may establish media sessions. SIP also supports multicast conferencing without requiring a separate conferencing unit. Basic conference management and statistics are provided by RTCP.

MCUs have been built to handle medium sized conferences but are not likely to be scalable to the size of conferences currently handled by multicast on the Internet today. While H.323 terminals can support multicast, MCUs are still a required element in the conference, and many of the MCU functions overlap with functions in RTCP [2].

Both SIP and H.323 utilize multicast for registration and server discovery.

8.4.8 Extensibility

H.323 can be extended to implement new features and functionality. Due to the ASN.1 PER encoding scheme and other constraints, however, extension "hooks" built into the protocol must be used to achieve this. If there is no place holder built in to the protocol in a particular place, then it cannot be extended there. Also, fields have strict limitations on content and length that support current features, which could limit future features.

In SIP, new methods, response codes, and headers can be easily added. Because unknown response codes can be interpreted by their class, and unknown headers can be ignored, some interoperability is possible even between SIP devices with different extensions implemented.

8.5 Comparison Summary

SIP and H.323 were developed by different organizations with different requirements and perspective on the world. It is likely that SIP and H.323 will co-exist for a number of years to come. Table 8.3 summarizes this comparison.

Table 8.3
SIP and H.323 Comparison Summary

Area	SIP	H.323
Transport	TCP, UDP, or other	TCP only
Conferencing	IP multicast	MCU required
Encoding	Text	Binary
Type of spec	Signaling only	Umbrella—covers all aspects
Vendor support	Growing	Strong
Features	PSTN and Internet	Mainly PSTN
Versions	Only one implemented	Three implemented (V1, V2, and V3)
Server types	Proxies, redirect, registrars—stateful or stateless	Gatekeeper—stateful only
Loop detection	Stateless	Stateful

References

[1] "Packet-based Multimedia Communications Systems," International Telecommunications Union Recommendation H.323.

[2] Rosenberg, J., and H. Schulzrinne, "A Comparison of SIP and H.323 for Internet Telephony," *Network and Operating System Support for Digital Audio and Video (NOSSDAV)*, Cambridge, UK, July 1998.

[3] Agrawal, H. "SIP-H.323 Interworking Requirements," IETF Internet-Draft, Work in Progress.

[4] Agrawal, H. "SIP-H.323 Interworking," IETF Internet-Draft, Work in Progress.

9

Call Flow Examples

In this chapter, many of the concepts and details presented in the preceding chapters will be illustrated with examples. Each example includes a call flow diagram, a discussion of the example, followed by the message details. Each message is labeled in the figure with a message number for easy reference. For more examples of the protocol, refer to the SIP specificaton [1] and the SIP call flows [2] documents.

The purpose of the examples in this chapter is to illustrate aspects of the SIP protocol. The interoperation scenarios with the PSTN and with a H.323 network are not intended to fully define the interworking or show a complete parameter mapping between the protocols. Likewise, simplifications such as minimal authentication and direct client-to-gateway messaging are used to make the examples more clear.

9.1   SIP Call with Authentication, Proxies, and Record-Route

Figure 9.1 shows a basic SIP call between two SIP user agents involving two proxy servers. Rather than perform a DNS query on the SIP URL of the called party, the calling SIP phone sends the INVITE request to a proxy server for address resolution. The proxy server requires authentication to perform this service and replies with a 407 Proxy Authorization Required response. Using the nonce from the challenge, the caller resends the INVITE

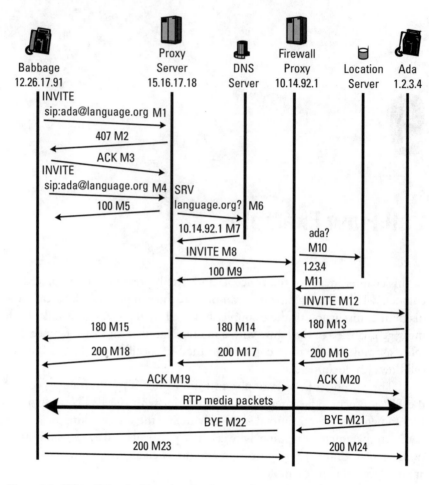

Figure 9.1 SIP-to-SIP call with authentication, proxies, and record-route.

with the caller's username and password credentials encrypted. The proxy checks the credentials, and finding them to be correct, performs the DNS lookup on the Request-URI. The INVITE is then forwarded to the proxy server listed in the DNS SRV record that handles the language.org domain. That proxy then looks up the Request-URI and locates a registration for the called party. The INVITE is forwarded to the destination UAS, a Record-Route header having been inserted to ensure that the proxy is present in all future requests by either party. This is because a direct routed SIP message to Ada would be blocked by the firewall.

The called party receives the INVITE request and sends 180 Ringing and 200 OK responses, which are routed back to the caller using the Via header chain from the initial INVITE. The ACK sent by the caller includes a Route header built from the Record-Route and Contact headers in the 200 OK response. This routing skips the first proxy but includes the firewall proxy. The media session begins with the user agents exchanging RTP and RTCP packets.

The call terminates when the called party, Ada, sends a BYE, which includes a Route header generated from the Record-Route header and the Contact header in the INVITE. Note that the CSeq for the called user agent is initialized to 1000. The acknowledgement of the BYE with a 200 OK response causes both sides to stop sending media packets.

```
M1    INVITE sip:ada@language.org SIP/2.0              ⇐Request-URI
      Via: SIP/2.0/UDP 12.26.17.91:5060
      From: Charles Babbage <sip:babbage@analyticalsoc.org>
      To: sip:ada@language.org
      Call-ID: f6-32-9a-34-91-e7@analyticalsoc.org
      CSeq: 1 INVITE                                   ⇐CSeq initialized to 1
      Contact: <sip:babbage@analyticalsoc.org>
      Subject: RE: Software
      User-Agent: Difference Engine 1
      Content-Type: application/sdp
      Content-Length: 137

      v=0
      o=Babbage 2890844534 2890844534 IN IP4 12.26.17.91
      s=-
      t=0 0
      c=IN IP4 12.26.17.91                             ⇐Babbage's IP address
      m=audio 49170 RTP/AVP 0                          ⇐Port number
      a=rtpmap:0 PCMU/8000                             ⇐Codec info


M2    SIP/2.0 407 Proxy Authentication Required
      Via: SIP/2.0/UDP 12.26.17.91:5060
      From: Charles Babbage <sip:babbage@analyticalsoc.org>
      To: sip:ada@language.org;tag=7360
      Call-ID: f6-32-9a-34-91-e7@analyticalsoc.org
      CSeq: 1 INVITE
      Proxy-Authenticate: Digest                       ⇐Authentication
        realm="SIP Telephone Company",                    challenge
        domain = sip-company.com ,
        nonce="9c8e88df84f1cec4341ae6e5a359",
        opaque="", stale="FALSE", algorithm="MD5"
```

```
M3    ACK sip:ada@language.org SIP/2.0
      Via: SIP/2.0/UDP 12.26.17.91:5060
      From: Charles Babbage <sip:babbage@analyticalsoc.org>
      To: sip:ada@language.org
      Call-ID: f6-32-9a-34-91-e7@analyticalsoc.org
      CSeq: 1 ACK                              ⇐CSeq not incremented
                                                 method set to ACK


M4    INVITE sip:ada@language.org SIP/2.0     ⇐INVITE resent with
      Via: SIP/2.0/UDP 12.26.17.91:5060                  credentials
      From: Charles Babbage <sip:babbage@analyticalsoc.org>
      To: sip:ada@language.org
      Call-ID:f6-32-9a-34-91-e7
                              @analyticalsoc.org    ⇐Call-ID unchanged
      CSeq: 2 INVITE                                ⇐CSeq incremented
      Proxy-Authorization:Digest
       username="Babbage",                          ⇐Credentials
       realm="SIP Telephone Company",
       nonce="9c8e88df84f1cec4341ae6e5a359",
       opaque="", uri="sip:proxy.sip-company.com",
       response="e56131d19580cd833064787ecc"
      Contact: sip:babbage@analyticalsoc.org
      Subject: RE: Software
      User-Agent: Difference Engine 1
      Content-Type: application/sdp
      Content-Length: 137

      v=0
      o=Babbage 2890844534 2890844534 IN IP4 12.26.17.91
      s=-
      t=0 0
      c=IN IP4 12.26.17.91
      m=audio 49170 RTP/AVP 0
      a=rtpmap:0 PCMU/8000


M5    SIP/2.0 100 Trying                       ⇐Provisional response
      Via: SIP/2.0/UDP 12.26.17.91:5060           indicates credentials
      From: Charles Babbage                          accepted by proxy
                     <sip:babbage@analyticalsoc.org>
      To: sip:ada@language.org
      Call-ID: f6-32-9a-34-91-e7@analyticalsoc.org
      CSeq: 2 INVITE


M6    DNS Query: SRV language.org?


M7    DNS SRV Record: 10.14.92.1
```

```
M8    INVITE sip:ada@language.org SIP/2.0
      Via: SIP/2.0/UDP 15.16.17.18:5060;branch=3f31049.1
      Via: SIP/2.0/UDP 12.26.17.91:5060
      From: Charles Babbage <sip:babbage@analyticalsoc.org>
      To: sip:ada@language.org
      Call-ID: f6-32-9a-34-91-e7@analyticalsoc.org
      CSeq: 2 INVITE
      Contact: sip:babbage@analyticalsoc.org
      Subject: RE: Software
      User-Agent: Difference Engine 1
      Content-Type: application/sdp
      Content-Length: 137

      v=0
      o=Babbage 2890844534 2890844534 IN IP4 12.26.17.91
      s=-
      t=0 0
      c=IN IP4 12.26.17.91
      m=audio 49170 RTP/AVP 0
      a=rtpmap:0 PCMU/8000
```

```
M9    SIP/2.0 100 Trying                         ⇐Provisional response
      Via:SIP/2.0/UDP 15.16.17.18:5060;                not forwarded
                      branch=3f31049.1                    by proxy
      Via: SIP/2.0/UDP 12.26.17.91:5060
      From: Charles Babbage <sip:babbage@analyticalsoc.org>
      To: sip:ada@language.org
      Call-ID: f6-32-9a-34-91-e7@analyticalsoc.org
      CSeq: 2 INVITE
```

```
M10   Location Service Query: ada?
```

```
M11   Location Service Response: 1.2.3.4
```

```
M12   INVITE sip:ada@1.2.3.4 SIP/2.0
      Via: SIP/2.0/UDP 10.14.92.1:5060;branch=24105.1
      Via: SIP/2.0/UDP 15.16.17.18:5060;branch=3f31049.1
      Via: SIP/2.0/UDP 12.26.17.91:5060
      From: Charles Babbage <sip:babbage@analyticalsoc.org>
      To: sip:ada@language.org
      Call-ID: f6-32-9a-34-91-e7@analyticalsoc.org
      CSeq: 2 INVITE
      Contact: sip:babbage@analyticalsoc.org
      Subject: RE: Software
      User-Agent: Difference Engine 1
      Record-Route: <sip:ada@language.org;          ⇐Record-route added
                     maddr=10.14.92.1>                    by proxy
```

```
Content-Type: application/sdp
Content-Length: 137

v=0
o=Babbage 2890844534 2890844534 IN IP4 12.26.17.91
s=-
t=0 0
c=IN IP4 12.26.17.91
m=audio 49170 RTP/AVP 0
a=rtpmap:0 PCMU/8000
```

M13 SIP/2.0 180 Ringing
```
Via: SIP/2.0/UDP 10.14.92.1:5060;branch=24105.1
Via: SIP/2.0/UDP 15.16.17.18:5060;branch=3f31049.1
Via: SIP/2.0/UDP 12.26.17.91:5060
From: Charles Babbage <sip:babbage@analyticalsoc.org>
To: <sip:ada@language.org>;tag=65a3547e3          ⇐Tag added by
Call-ID: f6-32-9a-34-91-e7@analyticalsoc.org        called party
CSeq: 2 INVITE
```

M14 SIP/2.0 180 Ringing
```
Via: SIP/2.0/UDP 15.16.17.18:5060;branch=3f31049.1
Via: SIP/2.0/UDP 12.26.17.91:5060
From: Charles Babbage <sip:babbage@analyticalsoc.org>
To: <sip:ada@language.org>;tag=65a3547e3
Call-ID: f6-32-9a-34-91-e7@analyticalsoc.org
CSeq: 2 INVITE
```

M15 SIP/2.0 180 Ringing
```
Via: SIP/2.0/UDP 12.26.17.91:5060
From: Charles Babbage <sip:babbage@analyticalsoc.org>
To: <sip:ada@language.org>;tag=65a3547e3
Call-ID: f6-32-9a-34-91-e7@analyticalsoc.org
CSeq: 2 INVITE
```

M16 SIP/2.0 200 OK   ⇐Call accepted
```
Via: SIP/2.0/UDP 10.14.92.1:5060;branch=24105.1
Via: SIP/2.0/UDP 15.16.17.18:5060;branch=3f31049.1
Via: SIP/2.0/UDP 12.26.17.91:5060
From: Charles Babbage <sip:babbage@analyticalsoc.org>
To: <sip:ada@language.org>;tag=65a3547e3
Call-ID: f6-32-9a-34-91-e7@analyticalsoc.org
CSeq: 2 INVITE
Contact: sip:ada@drawingroom.language.org
Record-Route: <sip:ada@language.org;                ⇐ Copied from
              maddr=10.14.92.1>                         INVITE
Content-Type: application/sdp
```

```
Content-Length: 126

v=0
o=Ada 2890844536 2890844536 IN IP4 1.2.3.4
s=-
t=0 0
c=IN IP4 1.2.3.4                              ⇐Ada's IP address
m=audio 52310 RTP/AVP 0                       ⇐Port number
a=rtpmap:0 PCMU/8000                          ⇐Codec information
```

M17
```
SIP/2.0 200 OK
Via: SIP/2.0/UDP 15.16.17.18:5060;branch=3f31049.1
Via: SIP/2.0/UDP 12.26.17.91:5060
From: Charles Babbage <sip:babbage@analyticalsoc.org>
To: <sip:ada@language.org>;tag=65a3547e3
Call-ID: f6-32-9a-34-91-e7@analyticalsoc.org
CSeq: 2 INVITE
Contact: sip:ada@drawingroom.language.org
Record-Route: <sip:ada@language.org;maddr=10.14.92.1>
Content-Type: application/sdp
Content-Length: 126

v=0
o=Ada 2890844536 2890844536 IN IP4 1.2.3.4
s=-
t=0 0
c=IN IP4 1.2.3.4
m=audio 52310 RTP/AVP 0
a=rtpmap:0 PCMU/8000
```

M18
```
SIP/2.0 200 OK
Via: SIP/2.0/UDP 12.26.17.91:5060
From: Charles Babbage <sip:babbage@analyticalsoc.org
To: <sip:ada@language.org>;tag=65a3547e3
Call-ID: f6-32-9a-34-91-e7@analyticalsoc.org
CSeq: 2 INVITE
Contact: sip:ada@drawingroom.language.org
Record-Route: <sip:ada@language.org;maddr=10.14.92.1>
Content-Type: application/sdp
Content-Length: 126

v=0
o=Ada 2890844536 2890844536 IN IP4 1.2.3.4
s=-
t=0 0
c=IN IP4 1.2.3.4
```

```
            m=audio 52310 RTP/AVP 0
            a=rtpmap:0 PCMU/8000
```

M19
```
      ACK sip:ada@drawingroom.language.org              ⇐Sent to firewall
                           SIP/2.0                          proxy due to
      Via: SIP/2.0/UDP 12.26.17.91:5060                    Record-Route
      From: Charles Babbage <sip:babbage@analyticalsoc.org> header
      To: <sip:ada@language.org>;tag=65a3547e3
      Call-ID: f6-32-9a-34-91-e7@analyticalsoc.org
      CSeq: 2 ACK
      Route: <sip:ada@drawingroom.language        ⇐Created from Record-
                           .org                      Route header
                                                     in 200 OK
```

M20
```
      ACK sip:ada@drawingroom.language.org SIP/2.0
      Via: SIP/2.0/UDP 10.14.92.1:5060;branch=24105.1
      Via: SIP/2.0/UDP 12.26.17.91:5060
      From: Charles Babbage <sip:babbage@analyticalsoc.org>
      To: <sip:ada@language.org>;tag=65a3547e3
      Call-ID: f6-32-9a-34-91-e7@analyticalsoc.org
      CSeq: 2 INVITE
```

M21
```
      BYE sip:babbage@analyticalsoc.org SIP/2.0
      Via: SIP/2.0/UDP 1.2.3.4:5060
      From: Ada Lovelace <sip:ada@language.org>;tag=65a3547e3
      To: Charles Babbage <sip:babbage@analyticalsoc.org>
      Call-ID: f6-32-9a-34-91-e7@analyticalsoc.org
      CSeq: 1000 BYE                           ⇐CSeq initialized to 1000
      Route: <sip:babbage@analyticsoc.org;
                         maddr=12.26.17.91>       ⇐From Record-
                                                   Route header
```

M22
```
      BYE sip:babbage@analyticalsoc.org SIP/2.0
      Via: SIP/2.0/UDP 10.14.92.1:5060;branch=24105.1
      Via: SIP/2.0/UDP 1.2.3.4:5060
      From: Ada Lovelace <sip:ada@language.org>;tag=65a3547e3
      To: Charles Babbage <sip:babbage@analyticalsoc.org>
      Call-ID: f6-32-9a-34-91-e7@analyticalsoc.org
      CSeq: 1000 BYE
```

M23
```
      SIP/2.0 200 OK
      Via: SIP/2.0/UDP 10.14.92.1:5060;branch=24105.1
      Via: SIP/2.0/UDP 1.2.3.4:5060
      From: Ada Lovelace <sip:ada@language.org>;tag=65a3547e3
      To: Charles Babbage <sip:babbage@analyticalsoc.org>
      Call-ID: f6-32-9a-34-91-e7@analyticalsoc.org
      CSeq: 1000 BYE
```

```
M24    SIP/2.0 200 OK
       Via: SIP/2.0/UDP 1.2.3.4:5060
       From: Ada Lovelace <sip:ada@language.org>;tag=65a3547e3
       To: Charles Babbage <sip:babbage@analyticalsoc.org>
       Call-ID: f6-32-9a-34-91-e7@analyticalsoc.org
       CSeq: 1000 BYE
```

9.2 SIP Call with Stateless and Stateful Proxies with Called Party Busy

Figure 9.2 shows an example of a SIP with a stateless proxy server and a stateful proxy server. The call is not completed because called party is busy. The called user agent initially sends a 180 Ringing response but then sends a 600 Busy Everywhere response containing a Retry-After header to indicate that the call is being rejected. The stateful proxy returns a 100 Trying response to the INVITE, and also acknowledges the 600 Busy Everywhere response with an ACK. The stateless proxy does not send a 100 Trying and forwards the 600 Busy Everywhere and the ACK sent by the caller user agent. Also note that the initial INVITE does not contain a message body.

```
M1     INVITE sip:schockley@transistor.org SIP/2.0
       Via: SIP/2.0/UDP discrete.sampling.org:5060
       From: Shannon <sip:shannon@sampling.org>;tag=1
```

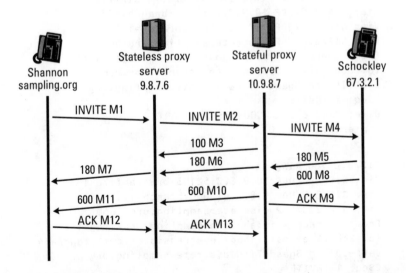

Figure 9.2 SIP call example with stateless and stateful proxies with busy called party.

```
         To: Schockley <sip:shockley@transistor.com>
         Call-ID: adf8gasdd7fld@discrete.sampling.org
         CSeq: 1 INVITE
         Date: Sat, 8 Jul 2000 08:23:00 GMT          ⇐Optional date header
         Content-Length: 0                           ⇐Optional Content-Length header


M2       INVITE sip:schockley@transistor.org SIP/2.0   ⇐Stateless proxy
         Via: SIP/2.0/UDP 9.8.7.6:5060;branch=1.1          does not send
         Via: SIP/2.0/UDP discrete.sampling.org:5060          100 Trying
         From: Shannon <sip:shannon@sampling.org>;tag=1
         To: Schockley <sip:shockley@transistor.com>
         Call-ID: adf8gasdd7fld@discrete.sampling.org
         CSeq: 1 INVITE
         Date: Sat, 8 Jul 2000 08:23:00 GMT
         Content-Length: 0


M3       SIP/2.0 100 Trying                          ⇐Stateful proxy does
         Via: SIP/2.0/UDP 9.8.7.6:5060;branch=1.1         send 100 Trying
         Via: SIP/2.0/UDP discrete.sampling.org:5060
         From: Shannon <sip:shannon@sampling.org>;tag=1
         To: Schockley <sip:shockley@transistor.com>
         Call-ID: adf8gasdd7fld@discrete.sampling.org
         CSeq: 1 INVITE
         Content-Length: 0


M4       INVITE sip:schockley@transistor.org SIP/2.0
         Via: SIP/2.0/UDP 10.9.8.7:52103;branch=ff7d.1
         Via: SIP/2.0/UDP 9.8.7.6:5060;branch=1.1
         Via: SIP/2.0/UDP discrete.sampling.org:5060
         From: Shannon <sip:shannon@sampling.org>;tag=1
         To: Schockley <sip:shockley@transistor.com>
         Call-ID: adf8gasdd7fld@discrete.sampling.org
         CSeq: 1 INVITE
         Date: Sat, 8 Jul 2000 08:23:00 GMT
         Content-Length: 0


M5       SIP/2.0 180 Ringing
         Via: SIP/2.0/UDP 10.9.8.7:52103;branch=ff7d.1
         Via: SIP/2.0/UDP 9.8.7.6:5060;branch=1.1
         Via: SIP/2.0/UDP discrete.sampling.org:5060
         From: Shannon <sip:shannon@sampling.org>;tag=1
         To: Schockley <sip:shockley@transistor.com>;tag=aa34
         Call-ID: adf8gasdd7fld@discrete.sampling.org
         CSeq: 1 INVITE
         Content-Length: 0
```

```
M6    SIP/2.0 180 Ringing
      Via: SIP/2.0/UDP 9.8.7.6:5060;branch=1.1
      Via: SIP/2.0/UDP discrete.sampling.org:5060
      From: Shannon <sip:shannon@sampling.org>;tag=1
      To: Schockley <sip:shockley@transistor.com>;tag=aa34
      Call-ID: adf8gasdd7fld@discrete.sampling.org
      CSeq: 1 INVITE
      Content-Length: 0


M7    SIP/2.0 180 Ringing
      Via: SIP/2.0/UDP discrete.sampling.org:5060
      From: Shannon <sip:shannon@sampling.org>;tag=1
      To: Schockley <sip:shockley@transistor.com>;tag=aa34
      Call-ID: adf8gasdd7fld@discrete.sampling.org
      CSeq: 1 INVITE
      Content-Length: 0


M8    SIP/2.0 600 Busy Everywhere                    ⇐Schockley is busy
      Via: SIP/2.0/UDP 10.9.8.7:52103;branch=ff7d.1
      Via: SIP/2.0/UDP 9.8.7.6:5060;branch=1.1
      Via: SIP/2.0/UDP discrete.sampling.org:5060
      From: Shannon <sip:shannon@sampling.org>;tag=1
      To: Schockley <sip:shockley@transistor.com>;tag=aa34
      Call-ID: adf8gasdd7fld@discrete.sampling.org
      CSeq: 1 INVITE
      Retry-After: Sun, 9 Jul 2000 11:59:00 GMT
      Content-Length: 0


M9    ACK sip:schockley@transistor.com SIP/2.0       ⇐Stateful proxy
      Via: SIP/2.0/UDP 10.9.8.7:52103;branch=5f7e.1    ACKs response
      From: Shannon <sip:shannon@sampling.org>;tag=1
      To: Schockley <sip:shockley@transistor.com>;tag=aa34
      Call-ID: adf8gasdd7fld@discrete.sampling.org
      CSeq: 1 ACK
      Content-Length: 0


M10   SIP/2.0 600 Busy Everywhere
      Via: SIP/2.0/UDP 9.8.7.6:5060;branch=1.1
      Via: SIP/2.0/UDP discrete.sampling.org:5060
      From: Shannon <sip:shannon@sampling.org>;tag=1
      To: Schockley <sip:shockley@transistor.com>;tag=429
      Call-ID: adf8gasdd7fld@discrete.sampling.org
      CSeq: 1 INVITE
      Retry-After: Sun, 9 Jul 2000 11:59:00 GMT
      Content-Length: 0
```

```
M11   SIP/2.0 600 Busy Everywhere                    ⇐Stateless proxy does
      Via: SIP/2.0/UDP discrete.sampling.org:5060      not ACK response
      From: Shannon <sip:shannon@sampling.org>;tag=1
      To: Schockley <sip:shockley@transistor.com>;tag=429
      Call-ID: adf8gasdd7fld@discrete.sampling.org
      CSeq: 1 INVITE
      Retry-After: Sun, 9 Jul 2000 11:59:00 GMT
      Content-Length: 0
```

```
M12   ACK sip:schockley@transistor.com SIP/2.0
      Via: SIP/2.0/UDP discrete.sampling.org:5060
      From: Shannon <sip:shannon@sampling.org>;tag=1
      To: Schockley <sip:shockley@transistor.com>;tag=429
      Call-ID: adf8gasdd7fld@discrete.sampling.org
      CSeq: 1 ACK
      Content-Length: 0
```

```
M13   ACK sip:schockley@transistor.com SIP/2.0
      Via: SIP/2.0/UDP 9.8.7.6:5060;branch=5.1
      Via: SIP/2.0/UDP discrete.sampling.org:5060
      From: Shannon <sip:shannon@sampling.org>;tag=1
      To: Schockley <sip:shockley@transistor.com>;tag=429
      Call-ID: adf8gasdd7fld@discrete.sampling.org
      CSeq: 1 ACK
      Content-Length: 0
```

9.3   SIP to PSTN Call Through Gateway

In the example shown in Figure 9.3, the calling SIP phone places a telephone call to the PSTN through a PSTN gateway. The SIP phone collects the dialed digits and puts them into a SIP URL used in the Request-URI and the To header. The caller may have dialed either the globalized phone number 1-202-555-1313 or they may have just dialed a local number 555-1313, and the SIP phone added the assumed country code and area code to produce the globalized URL. The SIP phone has been preconfigured with the IP address of the PSTN gateway, so it is able to send the INVITE directly to gw.carrier.com. The gateway initiates the call into the PSTN by selecting an SS7 ISUP trunk to the next telephone switch in the PSTN. The dialed digits from the INVITE are mapped into the ISUP Initial Address Message (IAM). The ISUP Address Complete Message (ACM) is sent back by the PSTN to indicate that the trunk has been seized. Progress tones are generated in the one-way audio path established in the PSTN. In this example, ring tone is generated by the far end telephone switch. The gateway maps the

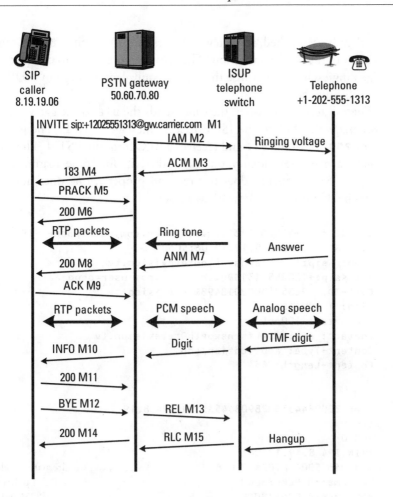

Figure 9.3 SIP to PSTN call through gateway.

ACM to the 183 Session Progress response containing SDP indicating the RTP port that the gateway will bridge the audio from the PSTN. Upon reception of the 183, the caller's UAC begins receiving the RTP packets sent from the gateway and presents the audio to the caller so they know that the call is progressing in the PSTN.

The call completes when the called party answers the telephone, which causes the telephone switch to send an Answer Message (ANM) to the gateway. The gateway then cuts the PSTN audio connection through in both directions and sends a 200 OK response to the caller. Because the RTP media path is already established, the gateway echoes the SDP in the 183 but causes no changes to the RTP connection. The UAC sends an ACK to complete the SIP signaling exchange. Because there is no equivalent message in ISUP, the gateway absorbs the ACK.

The call terminates when the caller sends the BYE to the gateway. The gateway maps the BYE to the ISUP Release message or REL. The gateway sends the 200 OK to the BYE and receives a RLC from the PSTN. These two messages have no dependency on each other; if, for some reason, either the SIP or PSTN network does not respond properly, one does not want resources held in the other network as a result.

```
M1    INVITE sip:+12025551313@gw.carrier.com;user=phone SIP/2.0
      Via: SIP/2.0/UDP 8.19.19.06:5060
      From: <sip:filo.farnsworth@television.tv>
      To: <sip:+12025551313@gw.carrier.com;user=phone>
      Call-ID: 49235243082018498@television.tv
      CSeq: 1 INVITE
      Supported: 100rel
      Contact: sip:filo.farnsworth@television.tv
      Content-Type: application/sdp
      Content-Length: 154

      v=0
      o=FF 2890844535 2890844535 IN IP4 8.19.19.06
      s=-
      t=0 0
      c=IN IP4 8.19.19.06
      m=audio 5004 RTP/AVP 0 8                    ⇐Two alternative codecs,
      a=rtpmap:0 PCMU/8000                              PCM μ-Law or
      a=rtpmap:8 PCMA/8000                                 PCM A-Law


M2    IAM
      CdPN=202-555-1313, NPI=E.164,
                          NOA=National        ⇐Gateway maps telephone
      USI=Speech                                  number from SIP URL
                                               into called party number


M3    ACM


M4    SIP/2.0 183 Session Progress
      Via: SIP/2.0/UDP 8.19.19.06:5060
      From: <sip:filo.farnsworth@television.tv>
      To: <+12025551313@gw.carrier.com;user=
                          phone>;tag=37         ⇐Tag and brackets added
```

```
Call-ID: 49235243082018498@television.tv
CSeq: 1 INVITE
RSeq: 08071
Content-Type: application/sdp
Content-Length: 139


v=0
o=Port1723 2890844535 2890844535 IN IP4 50.60.70.80
s=-
t=0 0
c=IN IP4 50.60.70.80
m=audio 62002 RTP/AVP 0                    ⇐Gateway selects M-Law codec
a=rtpmap:0 PCMU/8000
```

M5   ```
     PRACK sip:+12025551313@gw.carrier.com;user=phone SIP/2.0
     Via: SIP/2.0/UDP 8.19.19.06:5060
     From: <sip:filo.farnsworth@television.tv>;tag=37
     To: <sip:+12025551313@gw.carrier.com;user=phone>
     Call-ID: 49235243082018498@television.tv
     CSeq: 2 PRACK
     RAck: 08071 1 INVITE
     Content-Length: 0
     ```

M6   ```
     SIP/2.0 200 OK
     Via: SIP/2.0/UDP 8.19.19.06:5060
     From: <sip:filo.farnsworth@television.tv>;tag=37
     To: <sip:+12025551313@gw.carrier.com;user=phone>
     Call-ID: 49235243082018498@television.tv
     CSeq: 2 PRACK
     ```

M7   **ANM**

M8   ```
     SIP/2.0 200 OK
     Via: SIP/2.0/UDP 8.19.19.06:5060
     From: <sip:filo.farnsworth@television.tv>
     To: <+12025551313@gw.carrier.com;user=phone>;tag=37
     Call-ID: 49235243082018498@television.tv
     CSeq: 1 INVITE
     Content-Type: application/sdp
     Content-Length: 139

     v=0
     o=Port1723 2890844535 2890844535 IN IP4 50.60.70.80
     s=-
     t=0 0
     c=IN IP4 50.60.70.80
     ```
     ```

```
      m=audio 62002 RTP/AVP 0
      a=rtpmap:0 PCMU/8000
```

M9   ACK sip:+12025551313@gw.carrier.com;user=phone SIP/2.0
```
      Via: SIP/2.0/UDP 8.19.19.06:5060
      From: <sip:filo.farnsworth@television.tv>
      To: <+12025551313@gw.carrier.com;user=phone>;tag=37
      Call-ID: 49235243082018498@television.tv
      CSeq: 1 ACK
```

M10   INFO sip:filo.farnsworth@television.tv SIP/2.0
```
      Via: SIP/2.0/UDP 50.60.70.80:5060
      From: <sip:+12025551313@gw.carrier.com;user=phone>;tag=37
      To: <sip:filo.farnsworth@television.tv>
      Call-ID: 49235243082018498@television.tv
      CSeq: 1 INFO
      Content-Type: text/plain
      Content-Length: 10

      DTMF  "9"
```

M11   SIP/2.0 200 OK
```
      Via: SIP/2.0/UDP 50.60.70.80:5060
      From: <sip:+12025551313@gw.carrier.com;user=phone>;tag=37
      To: <sip:filo.farnsworth@television.tv>
      Call-ID: 49235243082018498@television.tv
      CSeq: 1 INFO
```

M12   BYE sip:+12025551313@gw.carrier.com;user=phone SIP/2.0
```
      Via: SIP/2.0/UDP 8.19.19.06:5060
      From: <sip:filo.farnsworth@television.tv>
      To: <+12025551313@gw.carrier.com;user=phone>;tag=37
      Call-ID: 49235243082018498@television.tv
      CSeq: 3 BYE                                    ⇐CSeq incremented
```

M13   REL
```
      CauseCode=16 Normal Clearing
```

M14   SIP/2.0 200 OK
```
      Via: SIP/2.0/UDP 8.19.19.06:5060
      From: <sip:filo.farnsworth@television.tv>
      To: <+12025551313@gw.carrier.com;user=phone>;tag=37
      Call-ID: 49235243082018498@television.tv
      CSeq: 3 BYE
```

M15   RLC

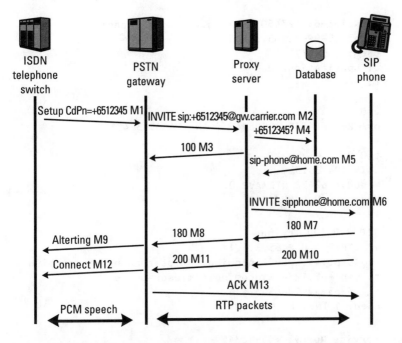

Figure 9.4 PSTN to SIP phone through gateway.

9.4 PSTN to SIP Call Through Gateway

Figure 9.4 shows a call originating from a telephone in the PSTN that terminates on a SIP phone in the Internet. The compact form of SIP is used throughout the example. Note that there is no compact form for `cseq` due to an oversight in the standard document.

```
M1    Setup
      CdPN=6512345, NPI=E.164,
                        NOA=International        ⇐Dialed telephone number
      CgPN=4567890, NPI=E.164,
                        NOA=International        ⇐PSTN caller's number
      USI=Speech

M2    INVITE sip:+6512345@incoming.com          ⇐Number mapped
                        ;user=phone SIP/2.0            into SIP URL
      v: SIP/2.0/UDP 65.3.4.1                   ⇐Compact form of headers
      f: <sip:+45.67890@incoming.com;user=phone>;   Caller includes tag
                        tag=6a589b1
```

```
        t: <sip:+65.12345@incoming.com;user=phone>
        i: a3-65-99-1d@65.3.4.1
        CSeq: 1 INVITE
        c: application/sdp
        l: 126

        v=0
        o=- 2890844535 2890844535 IN IP4 65.3.4.1
        s=-
        t=0 0
        c=IN IP4 65.3.4.1
        m=audio 62432 RTP/AVP 0
        a=rtpmap:0 PCMU/8000


M3      SIP/2.0 100 Trying
        v: SIP/2.0/UDP 65.3.4.1
        f: <sip:+45.67890@incoming.com;user=phone>;tag=6a589b1
        t: sip:+65.12345@incoming.com;user=phone
        i: a3-65-99-1d@65.3.4.1
        CSeq: 1 INVITE


M4      Service Query: +65-12345


M5      Location Service Response:
                          sip:user@home.com     ⇐Number maps to SIP URL


M6      INVITE sip:user@home.com SIP/2.0
        v: SIP/2.0/UDP 176.5.8.2:5060;branch=942834822.1
        v: SIP/2.0/UDP 65.3.4.1
        f: <sip:+45.67890@incoming.com;user=phone>;tag=6a589b1
        t: sip:+65.12345@incoming.com;user=phone
        i: a3-65-99-1d@65.3.4.1
        CSeq: 1 INVITE
        c: application/sdp
        l: 126

        v=0
        o=- 2890844535 2890844535 IN IP4 65.3.4.1
        s=-
        t=0 0
        c=IN IP4 65.3.4.1
        m=audio 62432 RTP/AVP 0
        a=rtpmap:0 PCMU/8000


M7      SIP/2.0 180 Ringing
        v: SIP/2.0/UDP 176.5.8.2:5060;branch=942834822.1
```

```
     v: SIP/2.0/UDP 65.3.4.1
     f: <sip:+45.67890@incoming.com;user=phone>;tag=6a589b1
     t: <sip:+65.12345@incoming.com;user=phone>;
                             tag=8657            ⇐Called party
     i: a3-65-99-1d@65.3.4.1                       adds tag


M8   SIP/2.0 180 Ringing
     v: SIP/2.0/UDP 65.3.4.1
     f: <sip:+45.67890@incoming.com;user=phone>;tag=6a589b1
     t: <sip:+65.12345@incoming.com;user=phone>;tag=8657
     i: a3-65-99-1d@65.3.4.1


M9   Alerting


M10  SIP/2.0 200 OK
     v: SIP/2.0/UDP 176.5.8.2:5060;branch=942834822.1
     v: SIP/2.0/UDP 65.3.4.1
     f: <sip:+45.67890@incoming.com;user=phone>;tag=6a589b1
     t: <sip:+65.12345@incoming.com;user=phone>;tag=8657
     i: a3-65-99-1d@65.3.4.1
     CSeq: 1 INVITE
     m: sip:user@home.com
     c: application/sdp
     l: 125

     v=0
     o=- 2890844565 2890844565 IN IP4 7.8.9.10
     s=-
     t=0 0
     c=IN IP4 7.8.9.10
     m=audio 5004 RTP/AVP 0
     a=rtpmap:0 PCMU/8000


M11  SIP/2.0 200 OK
     v: SIP/2.0/UDP 65.3.4.1
     f: <sip:+45.67890@incoming.com;user=phone>;tag=6a589b1
     t: <sip:+65.12345@incoming.com;user=phone>;tag=8657
     i: a3-65-99-1d@65.3.4.1
     CSeq: 1 INVITE
     m: sip:user@home.com
     c: application/sdp
     l: 125

     v=0
     o=- 2890844565 2890844565 IN IP4 7.8.9.10
     s=-
```

```
      t=0 0
      c=IN IP4 7.8.9.10
      m=audio 5004 RTP/AVP 0
      a=rtpmap:0 PCMU/8000
```

M12   `Connect`

M13   ```
      ACK sip:user@home.com SIP/2.0
      v: SIP/2.0/UDP 65.3.4.1
      f: <sip+45.67890@incoming.com;user=phone>;tag=6a589b1
      t: <sip+65.12345@incoming.com;user=phone>;tag=8657
      i: a3-65-99-1d@65.3.4.1
      CSeq: 1 ACK
      ```

## 9.5  Parallel Search

In this example the caller receives multiple possible locations for the called party from a redirect server. Instead of trying the locations one at a time, the user agent implements a parallel search for the called party by simultaneously sending the INVITE to three different locations, as shown in Figure 9.5. The SIP specification gives an example of this behavior in a proxy server, which is called a forking proxy [3].

In this example the first location responds with a 404 Not Found response. The second location responds with a 180 Ringing response, while the third location returns a 180 Ringing then a 200 OK response. The caller then sends an ACK to the third location to establish the call. Because one successful response has been received, a CANCEL is sent to the second location to terminate the search. The second location sends a 200 OK to the CANCEL and a 487 Request Cancelled to the INVITE.

This example shows some customized reason phrases in messages M7, M10, and M11.

M1    ```
      INVITE sip:faraday@effect.org;user=ip SIP/2.0
      Via: SIP/2.0/UDP 7.9.18.12:60000              ⇐Port 60000 is used
      From: J.C. Maxwell                               instead of 5060
        <sip:james.maxwell@kings.cambridge.edu.uk>    ⇐Long header
      To: <sip:faraday@effect.org;user=ip>              line wrapped
      Call-ID: mNjdwWjkBfWrd@7.9.18.12
      CSeq: 54 INVITE                               ⇐CSeq initialized to 54
      Contact:<sip:james.maxwell@kings.cambridge.edu.uk>
      Content-Type: application/sdp
      Content-Length: 129
      ```

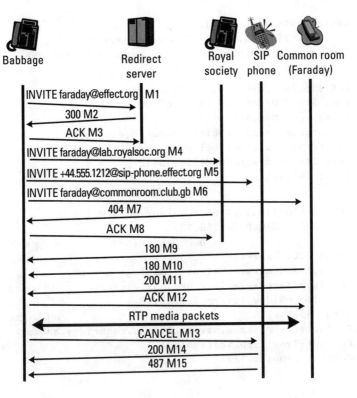

Figure 9.5 Parallel search example.

```
Content-Length: 129

v=0
o=max 2890844521 2890844521 IN IP4 7.9.18.12
s=-
t=0 0
c=IN IP4 7.9.18.12
m=audio 32166 RTP/AVP 4
a=rtpmap:4 DVI/8000
```

M2
```
SIP/2.0 300 Multiple locations            ⇐Redirect server returns
Via: SIP/2.0/UDP 7.9.18.12:60000                three locations
From: J.C. Maxwell
  <sip:james.maxwell@kings.cambridge.edu.uk>
To:<sip:faraday@effect.org;user=ip>;tag=1024
Call-ID: mNjdwWjkBfWrd@7.9.18.12
CSeq: 54 INVITE
```

```
Contact:<sip:faraday@lab.royalsoc.gb>
Contact:<sip:+44.555.1212@sip-phone.effect.org;user=phone>
Contact: <sip:michael.faraday@commonroom.club.gb>
```

M3   ACK sip:faraday@effect.org;user=ip
       Via: SIP/2.0/UDP 7.9.18.12:60000
       From: J.C. Maxwell
        <sip:james.maxwell@kings.cambridge.edu.uk>
       To: <sip:faraday@effect.org;user=ip>;tag=1024
       Call-ID: mNjdwWjkBfWrd@7.9.18.12
       CSeq: 54 INVITE

M4   INVITE sip:faraday@lab.royalsoc.gb SIP/2.0
       Via: SIP/2.0/UDP 7.9.18.12:
                              60000;branch=1   ⇐Each INVITE has
       From: J.C. Maxwell   unique branch ID
        <sip:james.maxwell@kings.cambridge.edu.uk>
       To: <sip:faraday@effect.org;user=ip>   ⇐Tag is not copied
       Call-ID: mNjdwWjkBfWrd@7.9.18.12   ⇐Call-ID unchanged
       CSeq: 55 INVITE   ⇐CSeq incremented
       Contact: <sip:james.maxwell@kings.cambridge.edu.uk>
       Content-Type: application/sdp
       Content-Length: 129

       v=0
       o=max 2890844521 2890844521 IN IP4 7.9.18.12
       s=-
       t=0 0
       c=IN IP4 7.9.18.12
       m=audio 32166 RTP/AVP 4
       a=rtpmap:4 DVI/8000

M5   INVITE sip:+44.555.1212@sip-phone.effect.org;
                              user=phone SIP/2.0
       Via: SIP/2.0/UDP 7.9.18.12:60000;branch=2
       From: J.C. Maxwell
        <sip:james.maxwell@kings.cambridge.edu.uk>
       To: <sip:faraday@effect.org;user=ip>
       Call-ID: mNjdwWjkBfWrd@7.9.18.12
       CSeq: 55 INVITE
       Contact: <sip:james.maxwell@kings.cambridge.edu.uk>
       Content-Type: application/sdp
       Content-Length: 129

       v=0
       o=max 2890844521 2890844521 IN IP4 7.9.18.12
```
```

```
s=-
t=0 0
c=IN IP4 7.9.18.12
m=audio 32166 RTP/AVP 4
a=rtpmap:4 DVI/8000
```

M6   INVITE sip:faraday@commonroom.club.gb SIP/2.0
     Via: SIP/2.0/UDP 7.9.18.12:60000;branch=3
     From: J.C. Maxwell
      <sip:james.maxwell@kings.cambridge.edu.uk>
     To: <sip:faraday@effect.org;user=ip>
     Call-ID: mNjdwWjkBfWrd@7.9.18.12
     CSeq: 55 INVITE
     Contact: <sip:james.maxwell@kings.cambridge.edu.uk>
     Content-Type: application/sdp
     Content-Length: 129

```
     v=0
     o=max 2890844521 2890844521 IN IP4 7.9.18.12
     s=-
     t=0 0
     c=IN IP4 7.9.18.12
     m=audio 32166 RTP/AVP 4
     a=rtpmap:4 DVI/8000
```

M7   SIP/2.0 404 The member you have requested is not available
     Via: SIP/2.0/UDP 7.9.18.12:60000;branch=1
     From: J.C. Maxwell
      <sip:james.maxwell@kings.cambridge.edu.uk>
     To: <sip:faraday@effect.org;user=ip>;tag=f6
     Call-ID: mNjdwWjkBfWrd@7.9.18.12
     CSeq: 55 INVITE

M8   ACK sip:faraday@lab.royalsoc.gb SIP/2.0
     Via: SIP/2.0/UDP 7.9.18.12:60000
     From: J.C. Maxwell <sip:james.maxwell@kings.cambridge.edu.uk>
     To: <sip:faraday@effect.org;user=ip>;tag=f6
     Call-ID: mNjdwWjkBfWrd@7.9.18.12
     CSeq: 55 ACK

M9   SIP/2.0 180 Ringing
     Via: SIP/2.0/UDP 7.9.18.12:60000;branch=2
     From: J.C. Maxwell <sip:james.maxwell@kings.cambridge.edu.uk>
     To: <sip:faraday@effect.org;user=ip>;tag=6321
     Call-ID: mNjdwWjkBfWrd@7.9.18.12
     CSeq: 55 INVITE

M10 SIP/2.0 180 Please wait while we locate Mr. Faraday
     Via: SIP/2.0/UDP 7.9.18.12:60000;branch=3
     From: J.C. Maxwell <sip:james.maxwell@kings.cambridge.edu.uk>
     To: <sip:faraday@effect.org;user=ip>;tag=531
     Call-ID: mNjdwWjkBfWrd@7.9.18.12
     CSeq: 55 INVITE

M11 SIP/2.0 200 Mr. Faraday at your service?
     Via: SIP/2.0/UDP 7.9.18.12:60000;branch=3
     From: J.C. Maxwell
       <sip:james.maxwell@kings.cambridge.edu.uk>
     To: <sip:faraday@effect.org;user=ip>;tag=531
     Call-ID: mNjdwWjkBfWrd@7.9.18.12
     CSeq: 55 INVITE
     User-Agent: PDV v4
     Contact: <sip:faraday@commonroom.club.gb>
     Content-Type: application/sdp
     Content-Length: 131

     v=0
     o=max 2890844521 2890844521 IN IP4 6.22.17.89
     t=0 0
     c=IN IP4 6.22.17.89
     m=audio 43782 RTP/AVP 4
     a=rtpmap:4 DVI/8000

M12 ACK sip:faraday@commonroom.club.gb;user=ip SIP/2.0
     Via: SIP/2.0/UDP 7.9.18.12:60000;branch=3
     From: J.C. Maxwell
       <sip:james.maxwell@kings.cambridge.edu.uk>
     To: <sip:faraday@effect.org;user=ip>;tag=531
     Call-ID: mNjdwWjkBfWrd@7.9.18.12
     CSeq: 55 ACK

M13 CANCEL sip:+44.555.1212@sip-phone.effect.org;
                      user=phone SIP/2.0
     Via: SIP/2.0/UDP 7.9.18.12:60000;branch=2   ⇐Cancels search
     From: J.C. Maxwell
       <sip:james.maxwell@kings.cambridge.edu.uk>
     To: <sip:faraday@effect.org;user=ip>;tag=6321
     Call-ID: mNjdwWjkBfWrd@7.9.18.12
     CSeq: 55 CANCEL   ⇐ CSeq not incremented,
                                          method set to CANCEL
M14 SIP/2.0 200 OK   ⇐CANCEL acknowledged
     Via: SIP/2.0/UDP 7.9.18.12:60000;branch=2

```
From: J.C. Maxwell <sip:james.maxwell@kings.cambridge.edu.uk>
To: <sip:faraday@effect.org;user=ip>;tag=6321
Call-ID: mNjdwWjkBfWrda7.9.18.12
CSeq: 55 CANCEL
```

M15 SIP/2.0 487 Request Cancelled   ⇐Final response to INVITE
```
     Via: SIP/2.0/UDP 7.9.18.12:60000;branch=2
     From: J.C. Maxwell <sip:james.maxwell@kings.cambridge.edu.uk>
     To: <sip:faraday@effect.org;user=ip;tag=6321>
     Call-ID: mNjdwWjkBfWrda7.9.18.12
     CSeq: 55 INVITE
```

9.6 H.323 to SIP Call

In this example, a H.323 terminal calls a SIP-enabled PC through a H.323/SIP gateway. The gateway does signaling translation between the protocols but allows the two end-points to exchange media packets directly with each other. The full details of SIP/H.323 interworking are being developed in the SIP working group [4].

In this example, shown in Figure 9.6, the initial message exchange is between the calling H.323 terminal and the H.323 gatekeeper. The gatekeeper resolves the H.323 alias into an address served by the H.323/SIP gateway. The ACF response indicates that gatekeeper-routed signaling is required, so the Q.931 and H.245 TCP connections are opened to the gatekeeper, which opens TCP connections to the gateway. The calling H.323 terminal sends a Q.931 Setup message to the gatekeeper, which proxies it to the H.323/SIP gateway. The gateway then looks up the H.323 alias and resolves it to the SIP URL of the called party. It constructs an INVITE from the Setup message and forwards it to a SIP proxy, which forwards it to the called party. Note that because the Setup message does not contain any media information, the INVITE does not contain any media information either. The called party sends a 180 Ringing then a 200 OK to indicate that the call has been answered. The media information present in the SDP message body is stored by the gateway, which sends Alerting and Connect messages to the gatekeeper, which proxies them to the calling H.323 terminal. The gateway holds off sending the ACK response to the INVITE until the H.245 media exchange is completed between the H.323 terminal and the gateway. Once that is complete, the negotiated media capabilities are returned in the ACK and the media session begins.

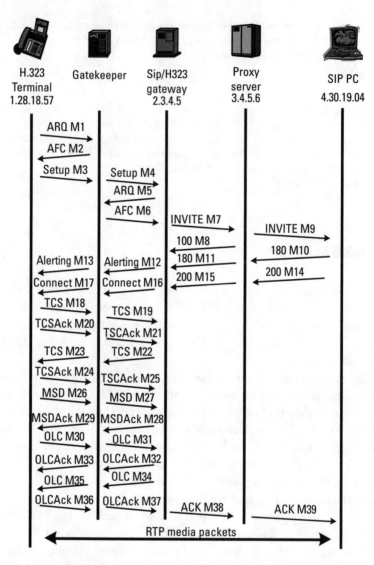

Figure 9.6 H.323 to SIP call.

M1   **ARQ**
      `address(h323alias=Stibitz)`

M2   **ACF**
      `gatekeeper routed signaling`

M3 Setup
    Cd address(h323alias=Stibitz)
    Cg address(h323alias=Burroughs)

M4 Setup
    Cd address(h323alias=Stibitz)
    Cg address(h323alias=Burroughs)

M5 ARQ

M6 ACF

M7 INVITE sip:stibitz@proxy.com SIP/2.0
    Via: SIP/2.0/TCP 2.3.4.5   ⇐TCP used for transport
    From: <sip:burroughs@h323-gateway.com>
    To: <sip:stibitz@proxy.com>
    Call-ID: 526272332146783569054
    CSeq: 43252 INVITE   ⇐CSeq initialized
    Contact: sip:burroughs@h323-gateway.com   to 43252

M8 100 Trying
    Via: SIP/2.0/TCP 2.3.4.5
    From: <sip:burroughs@h323-gateway.com>
    To: <sip:stibitz@proxy.com>
    Call-ID: 526272332146783569054
    CSeq: 43252 INVITE
    Content-Length: 0

M9 INVITE sip:gstibitz123@snailmail.com SIP/2.0
    Via: SIP/2.0/UDP 3.4.5.6:5060;
                        branch=452.1   ⇐Proxy forwards with UDP
    Via: SIP/2.0/TCP 2.3.4.5
    From: <sip:burroughs@h323-gateway.com>
    To: <sip:stibitz@proxy.com>
    Call-ID: 526272332146783569054
    CSeq: 43252 INVITE
    Contact: sip:burroughs@h323-gateway.com

M10 SIP/2.0 180 Ringing
    Via: SIP/2.0/UDP 3.4.5.6:5060; branch=452.1
    Via: SIP/2.0/TCP 2.3.4.5
    From: <sip:burroughs@h323-gateway.com>
    To: <sip:stibitz@proxy.com>;tag=1926

```
     Call-ID: 526272332146783569054
     CSeq: 43252 INVITE
     Content-Length: 0
```

M11 ```
     SIP/2.0 180 Ringing
     Via: SIP/2.0/TCP 2.3.4.5
     From: <sip:burroughs@h323-gateway.com>
     To: <sip:stibitz@proxy.com>;tag=1926
     Call-ID: 526272332146783569054
     CSeq: 43252 INVITE
     Content-Length: 0
```

M12  ```
     Alerting
```

M13 ```
     Alerting
```

M14  ```
     SIP/2.0 200 OK
     Via: SIP/2.0/UDP 3.4.5.6:5060
     Via: SIP/2.0/TCP 2.3.4.5
     From: <sip:burroughs@h323-gateway.com>
     To: <sip:stibitz@proxy.com>;tag=1926
     Call-ID: 526272332146783569054
     CSeq: 43252 INVITE
     Contact: <sip:gstibitz123@snailmail.com>
     Content-Type: application/sdp
     Content-Length: 134

     v=0
     o=George 2890844576 2890844576 IN IP4 4.30.19.04
     s=-
     t=0 0
     c=IN IP4 4.30.19.04
     m=audio 5004 RTP/AVP 0
     a=rtpmap:0 PCMU/8000
```

M15 ```
     SIP/2.0 200 OK
     Via: SIP/2.0/TCP 2.3.4.5
     From: <sip:burroughs@h323-gateway.com>
     To: <sip:stibitz@proxy.com>;tag=1926
     Call-ID: 526272332146783569054
     CSeq: 43252 INVITE
     Contact: <sip:gstibitz123@snailmail.com>
     Content-Type: application/sdp
     Content-Length: 134

     v=0
     o=George 2890844576 2890844576 IN IP4 4.30.19.04
```
```

```
s=-
t=0 0
c=IN IP4 4.30.19.04
m=audio 5004 RTP/AVP 0
a=rtpmap:0 PCMU/8000
Content-Length:0
```

M16   Connect

M17   Connect

M18   TerminalCapabilitySet

M19   TerminalCapabilitySet

M20   TerminalCapabilitySetAck

M21   TerminalCapabilitySetAck

M22   TerminalCapabilitySet

M23   TerminalCapabilitySet

M24   TerminalCapabilitySetAck

M25   TerminalCapabilitySetAck

M26   MasterSlaveDetermination

M27   MasterSlaveDetermination

M28   MasterSlaveDeterminationAck

M29   MasterSlaveDeterminationAck

M30   OpenLogicalChannel
      g711uLaw 1.28.18.57 60002

M31   OpenLogicalChannel
      g711uLaw 1.28.18.57 60002

M32   OpenLogicalChannelAck

M33   OpenLogicalChannelAck

```
M34    OpenLogicalChannel
       g711uLaw 4.30.19.04 5004


M35    OpenLogicalChannel
       g711uLaw 4.30.19.04 5004


M36    OpenLogicalChannelAck


M37    OpenLogicalChannelAck


M38    ACK sip:gstibitz123@snailmail.com SIP/2.0
       Via: SIP/2.0/TCP 2.3.4.5
       From: <sip:burroughs@h323-gateway.com>
       To: <sip:stibitz@proxy.com>;tag=1926
       Call-ID: 526272332146783569054
       CSeq: 43252 ACK
       Content-Type: application/sdp
       Content-Length: 130

       v=0
       o=- 2890844577 2890844577 IN IP4 1.28.18.57
       s=-
       t=0 0
       c=IN IP4 1.28.18.57
       m=audio 60002 RTP/AVP 0
       a=rtpmap:0 PCMU/8000


M39    ACK sip:gstibitz123@snailmail.com SIP/2.0
       Via: SIP/2.0/UDP 3.4.5.6:5060;branch=452.1
       Via: SIP/2.0/TCP 2.3.4.5
       From: <sip:burroughs@h323-gateway.com>
       To: <sip:stibitz@proxy.com>;tag=1926
       Call-ID: 526272332146783569054
       CSeq: 43252 ACK
       Content-Type: application/sdp
       Content-Length: 130

       v=0
       o=- 2890844577 2890844577 IN IP4 1.28.18.57
       s=-
       t=0 0
       c=IN IP4 1.28.18.57
       m=audio 60002 RTP/AVP 0
       a=rtpmap:0 PCMU/8000
```

References

[1] Handley, M., et al., "SIP: Session Initiation Protocol," RFC 2543, 1999, Section 16.

[2] Johnston, A., et al., "SIP Telephony Call Flow Examples," IETF Internet-Draft, Work in Progress.

[3] Handley, M., et al., "SIP: Session Initiation Protocol," RFC 2543, 1999, Forking Proxy Example.

[4] Agrawl, H., R. Roy, and V. Palawat, "SIP-H.323 Interworking Requirements," IETF Internet-Draft, Work in Progress.

10

Future Directions

This chapter discusses some of the current work items and design teams of the SIP working group as of July 2000, just prior to the IETF 48th Meeting. The extensions discussed here are currently only in Internet-Draft form, the precursor to the Request for Comments. For the latest on these extensions, check the IETF web site. Because Internet-Drafts are archived for only 6 months, some of the drafts mentioned in this chapter may no longer exist or may have a different name.

10.1   Changes to RFC 2543

Since the SIP protocol RFC 2543 was published, the SIP working group has maintained a list of bug fixes, clarifications, typos, and other minor modifications [1]. As of this writing, there is a document containing these modifications, published as an Internet-Draft in 2000. The chartered goal of the working group is to move SIP from its current Proposed Standard status to Draft Standard status. As a result, only minor changes will be made to the protocol, or else it will have to recycle back to Proposed Standard status and begin the standards process again. Most of the extensions that are working group items and that are close to RFC status have been described in previous chapters. These new methods, headers, and response codes are summarized in the following three tables. Table 10.1 lists the new proposed SIP methods

Table 10.1
Proposed SIP Methods

Method name	Description
INFO	Mid-call signaling information exchange (Section 4.1.7) [2]
PRACK	Acknowledgment of provisional responses (Section 4.1.8) [3]
REFER	Call transfer [4]
SUBSCRIBE	Request Notification of Call Event [5]
NOTIFY	Notification of Subscribed Call Event [5]
COMET	Call Preconditions Met [6]

or request message types. The first two, INFO and PRACK, are covered in Sections 4.1.7 and 4.1.8 and are close to RFC status. Two other methods, REFER and COMET are still being discussed and may change form before being standardized. SUBSCRIBE and NOTIFY have been developed in a related IETF working group called PINT. Table 10.2 lists the proposed new headers, all of which are covered in Chapter 7. Table 10.3 lists the two new proposed response codes, 183 and 421, both of which are covered in Chapter 5.

These new extensions to SIP will be documented in stand-alone RFCs. Future revisions to the RFC 2543 standard will roll these RFC documents into a single RFC.

10.2   SIP Working Group Design Teams

Design teams in the IETF are loosely organized voluntary groups of individuals working towards a common goal, usually an Internet-Draft document, that is of interest to the wider working group. In short, a design team is like a working group but on a much smaller scale. There are design teams working within the SIP IETF working group on the following areas: call control, convergence with PacketCable DCS extensions, call flows, SIP/H.323 interworking, home extension, SIP security, and SIP telephony. New design teams are constantly being formed as issues are raised at IETF meetings and from the mailing list.

Table 10.2
Proposed SIP Headers

Header Name	Description	Section
Accept-Contact	Caller preference for allowed URIs for proxy operation [7]	6.2.2
RAck	Reliable provisional response acknowledgement number [3]	6.2.18
Reject-Contact	Caller preference for rejected URIs for proxy operation [7]	6.2.13
Request-Disposition	Caller preference for type of server operation [7]	6.2.14
RSeq	Reliable provisional responses sequence number [3]	6.3.6
Session-Expires	Time limit on media session using Session Timer [8]	6.2.19
Supported	Lists options implemented by a User Agent or server [9]	6.1.10
Refer-To	URL for attended transfer [4]	

Table 10.3
Proposed SIP Response Codes

Response code	Description
183 Session Progress	Carries message body prior to final response [10]
421 Extension Required	A proxy to requires an extension [9]

10.2.1 Call Control

This chartered work item is to extend the SIP protocol to enable call controls such as call transfer and attended transfer, common features in the PSTN. Because the design of SIP is for end-device control rather than third-party call control, these extensions are not trivial. The initial work in this area defined new headers such as Bye-Also to initiate blind transfers. Blind transfers, also called unattended transfers, do not allow the party performing the transfer operation any way of knowing if the transfer succeeded or failed. A new approach, however, has been taken lately to instead define a new method REFER [4] that allows both unattended and attended transfer operations.

10.2.2 Convergence with PacketCable Distributed Call Signaling (DCS) Extensions

This work item involves the standardization of some of the extensions proposed to SIP by the PacketCable DCS group. This group led by the Cable-Labs consortium [10] has developed a set of protocol extensions for their application in providing home telephone service using the Data over Cable Service Interface Specification (DOCSIS) [11]. In the future, some of these proprietary extensions may be adopted into the SIP specification. Some of the extensions relate to billing, caller privacy, INVITE with no ring, QoS setup, etc. For example, COMET is defined in a DCS draft [6].

10.2.3 Call Flows

This work item involves the preparation of an informational RFC containing examples of the SIP protocol. The current draft [12] includes examples of the protocol's use both in an all IP environment as well as with gateways to the PSTN. The draft also includes example SIP messages developed for testing during the bakeoffs. A future Internet-Draft will include the use of the REFER method to implement common PSTN features.

10.2.4 SIP/H.323 Interworking

This working group design team is developing requirements [13] and recommendations for a standard interworking function for the signaling conversion between SIP and H.323. This design team is composed of experts on SIP from the IETF as well as experts on H.323 from the ITU Study Group 16.

10.2.5 Home Extension

This working group design team is developing call flows and implementation recommendations for how to implement home extensions to mimic current analog telephone behavior, as described in Section 3.8.

10.2.6 SIP Security

This working group design team is investigating security for SIP beyond the current use of SIP Digest and IPSec. The issue of security encompasses areas such as encryption, privacy of addresses, URLs, and IP addresses.

10.2.7 SIP for Telephony

This working group design team works on interworking and encapsulation of PSTN protocols with SIP. The SIP-T document [14] (formerly called "SIP Telephony Best Current Practices," or "SIP+") is an umbrella document that describes the use of SIP for communication between media gateways (PSTN to SIP to PSTN calls only) to achieve feature transparency to the PSTN devices. It specifies the use of:

- SIP version 2.0 in RFC 2543;
- MIME multipart specification [15];
- ISUP MIME type [16];
- INFO method [2];
- SIP/ISUP mapping draft [17];
- Network address point (NAP) to be defined in a future draft.

10.3 Other Related Drafts

Some other work items in the SIP working group relate to the SIP protocol without introducing new behavior or elements to the protocol. For example, the SIP Management Information Base (MIB) chartered work item has defined a standard MIB [18] for use by SNMP network management systems to control and manage SIP devices including user agents and servers. The DHCP SIP Option draft [19] specifies an extension to DHCP (Section 3.8) to allow a SIP user agent to locate a SIP server using DHCP. The SIP extension framework document [20] describes how the protocol should be extended. SIP will likely make use of the call processing language (CPL) [21] for uploading feature scripts being developed by the IPTEL working group.

References

[1]   Henning Schulzrinne has maintained this list of fixes on his excellent SIP web page at http://www.cs.columbia.edu/sip/. Information about how to subscribe to the SIP working group e-mail list or how to read the list archive is available on the SIP working group official charter web page at http://www.ietf.org/html.charters/sip-charter.html.

[2]   Donovan, S., "The SIP INFO Method," RFC 2976, 2000.

[3] Rosenberg, J., and H. Schulzrinne, "Reliability of Provisional Responses," IETF Internet-Draft, Work in Progress.

[4] Sparks, R., "SIP Call Control: REFER," IETF Internet-Draft, Work in Progress.

[5] Petrack S., and L. Genroy, "The PINT Service Protocol," RFC 2848, 2000.

[6] Marshall, W., et al., "Architectural Considerations for Providing Carrier Class Telephony Services Utilizing SIP-based Distributed Call Ceontrol Mechanisms," IETF Internet-Draft, Work in Progress.

[7] Schulzrinne, H., and J. Rosenberg, "SIP Caller Preferences and Callee Capabilities," IETF Internet-Draft, Work in Progress.

[8] Donovan, S., and J. Rosenberg, "The SIP Session Timer," IETF Internet-Draft, Work in Progress.

[9] Rosenberg, J., and H. Schulzrinne, "The SIP Supported Header," IETF Internet-Draft, Work in Progress.

[10] Donovan, S., et al., "SIP 183 Session Progress Message," IETF Internet-Draft, Work in Progress.

[11] Information about PacketCable and DOCSIS is available at http://www.cablelabs.com.

[12] Johnston, A., et al., "SIP Telephony Call Flow Examples," IETF Internet-Draft, Work in Progress.

[13] Agrawl, H., R. Roy, and V. Palawat, "SIP-H.323 Interworking Requirements," IETF Internet-Draft, Work in Progress.

[14] Zimmer, E., et al., "MIME Media Types for ISUP and QSIG Objects," IETF Internet-Draft, Work in Progress.

[15] Levinson, E., "The MIME Multipart/Related Content Type," RFC 2387

[16] Zimmer, E., "SIP Best Current Practice for Telephony Interworking," IETF Internet-Draft, Work in Progress.

[17] Camarillo, G., and A. Roach, "Best Current Practice for ISUP to SIP Mapping," IETF Internet-Draft, Work in Progress.

[18] Lingle, K., J. Maeng, and D. Walker, "Management Information Base for Session Initiation Protocol," IETF Internet-Draft, Work in Progress.

[19] Nair, G., and H. Schulzrinne, "DHCP Option for SIP Servers," IETF Internet-Draft, Work in Progress.

[20] Rosenberg, J., and H. Schulzrinne, "Guidelines for Authors of SIP Extensions," IETF Internet-Draft, Work in Progress.

[21] Lennox, J., and H. Schulzrinne, "CPL: A Language for User Control of Interent Telephony Services," IETF Internet-Draft, Work in Progress.

About the Author

Alan B. Johnston is a Distinguished Technical Member with WorldCom, Inc. in their Network Engineering Department, and is also an adjuct assistant professor of electrical engineering at Washington University in St. Louis. He is currently working with the SIP protcol in WorldCom's IP Communications Services project. Prior to WorldCom, he worked at a competitive local exchange carrier (CLEC), Brooks Fiber Properties, Telcordia (formerly Bellcore), and SBC Technology Resources. Alan is an active participant in the IETF SIP working group, and is the editor of the SIP IP Telephony Call Flows document, to become an informational RFC in 2001. *SIP: Understanding the Session Initiation Protocol* is his first book. Alan has a Ph.D. from Lehigh University, Bethlehem, Pennsylvania, in electrical engineering and a Bachelors of Engineering (First Class Honors) from the University of Melbourne, Melbourne, Australia in electrical engineering. Born in Melbourne, Australia, Alan currently resides in a national historic district in St. Louis, Missouri, with his wife Lisa, son Aidan, and daughter Nora.

Index

Recent Titles in the Artech House Telecommunications Library

Vinton G. Cerf, Senior Series Editor

For further information on these and other Artech House titles,
including previously considered out-of-print books now available
through our In-Print-Forever® (IPF®) program, contact:

Artech House	Artech House
685 Canton Street	46 Gillingham Street
Norwood, MA 02062	London SW1V 1AH UK
Phone: 781-769-9750	Phone: +44 (0)20 7596-8750
Fax: 781-769-6334	Fax: +44 (0)20 7630-0166
e-mail: artech@artechhouse.com	e-mail: artech-uk@artechhouse.com

Find us on the World Wide Web at:
www.artechhouse.com

o Wallstreet subscription

o PAS Blueprint presentation
  - BP 1.0 summary (of PAS)
  - talk to David
  - talk to Stephen

  - new concerns, models, rules
  - scope = Web
  - search the Web
  - general - meet or exceed customer expectation
           - beat competition in Web business
              - Schwab, ETrade ... (Yahoo?)
              - NB equiv - ?   ?)
           - general UI standards
              - different standards for different
                business functions

?

o distribution of RL Response 411, 030221 | 030305)
o ATP RFC 1889, SDP RFC 2327

Fidelity eBusiness

- search the web for (performance (web, requirements

- BB 1.0 summery in PAS area
- distribution of R (Respone, 411, 0302 21, 030305)